# The Himalayan Glaciers, Climate Change and Society

R. K. Ganjoo · M. N. Koul

# The Himalayan Glaciers, Climate Change and Society

## A Case Study of the Northwestern Himalayas, India

Springer

R. K. Ganjoo
Department of Geology
University of Jammu
Jammu, Jammu and Kashmir, India

M. N. Koul
Department of Geography
University of Jammu
Jammu, Jammu and Kashmir, India

ISBN 978-3-031-78282-4 ISBN 978-3-031-78283-1 (eBook)
https://doi.org/10.1007/978-3-031-78283-1



This Springer imprint is published by the registered company Springer Nature Switzerland AG
The registered company address is: Gewerbestrasse 11, 6330 Cham, Switzerland

# Acknowledgements

The book entitled *The Himalayan Glaciers, Climate Change and Society—A Case Study of the Northwestern Himalayas, India* is an outcome of the scientific work, spanning three decades, in the inhospitable terrain of Himalaya and braving the vagaries of unpredictable weather. The mission to explore the scientific truth of the interrelationship between Himalayan glaciers and global climate change could not have been successful without the support of human and financial resources. The authors would take the opportunity to acknowledge all the personnel and institutes for the support and financial assistance that was received at various stages of the scientific work.

The fleet of young researchers who joined the authors during the journey of three decades in achieving the scientific truth deserves a special mention here. It is the hard and sincere field input of young researchers that helped in building a robust scientific data bank. The young researchers who joined the scientific project at the initial stage were Dr. A. K. Raina and Sh. Ravi Sharma. The scientific team was subsequently joined by more young researchers, namely Dr. R. P. Sharma, Dr. Shyamal Sarkar, Dr. Ramesh Raina, Dr. Surinder Singh, Dr. Bhupinder Singh Sambyal and Dr. Ravi Kant. The authors could not have conceived the compilation of scientific work into the book without the dedicated work by the young researchers in the field and laboratory.

It is also a fact that continuation of scientific exploration in Himalayan glaciers for a long period of three decades could not have reached a conclusive and logical end unless there was regular source of financial support. The authors would like to put on record the adequate financial assistance received in the project mode at various stages of the scientific mission from Department of Science and Technology (Ministry of Science and Technology, Government of India) and Space Applications Centre (Indian Space Research Organization, Government of India). Besides the financial support, the authors are also grateful to various experts with the Department of Science and Technology and Space Applications Centre for their scientific deliberations to enhance the quality of our scientific work. The noteworthy experts to mention here are Sh. V. K. Raina, Sh. Deepak Srivastava, Dr. Ashwagosha Ganju, Dr. Ajai, Dr. I. M. Bahuguna, Dr. A. V. Kulkarni and Dr. A. K. Sharma. A very special thanks is to Sh. V. K. Raina (Former Deputy Director General, Geological Survey of

India) who enriched us with his vast knowledge on Himalayan glaciers and joined us for field exploration to share his expertise in the field. The authors are highly indebted to Late Prof. S. N. Rajaguru (Deccan College PGRI, Pune) for sharing his expertise on Quaternary geology and providing field training in Quaternary geology to one of the authors (RKG). The authors are also thankful to our contemporaries at various Indian universities and institutes who shared their scientific work and experience with us to help fine tune our research work. The fellow workers to mention here are Prof. Shakil K. Romshoo (University of Kashmir, Kashmir), Prof. Milap C. Sharma (Jawaharlal Nehru University, New Delhi), Prof. R. Bali (University of Lucknow, Lucknow), Prof. H. C. Nainwal (Hemavati Nandan Bahuguna University, Garhwal) and Dr. Kireet Kumar (G. B. Pant National Institute of Himalayan Environment and Development, Almora).

The authors shall fail in their duty if the administrative support received from the University of Jammu is not acknowledged here. It was because of the encouragement we received from our respective departments that we were able to carry out long (three to four months) field explorations in treacherous terrain. The infrastructure facility provided by the University of Jammu also receives appreciation and acknowledgement from the authors.

Last but not the least, thanks are to our life partners, namely Mrs. Sneh Ganjoo and Mrs. Santosh Koul, who encouraged, appreciated and shared our responsibilities to allow us to devote and dedicate our time to the scientific work and achieve the goal.

R. K. Ganjoo
M. N. Koul

# Contents

# About the Authors

**Professor R. K. Ganjoo** born in April 1959, obtained his Master's degree in Geology from University of Jammu and doctoral degree in Quaternary Geology from University of Poona (now Savitribai Phule Pune University). At the early stage of his academic career, he joined as Scientific Assistant at Deccan College Postgraduate and Research Institute, Pune, India (Deemed to be University). He worked in parts of Maharashtra, Madhya Pradesh and Rajasthan states in India and made a humble contribution in the field of fluvial and aeolian geomorphology. Subsequently, he joined the teaching faculty at University of Jammu (Jammu and Kashmir, India) in 1992. Prof. Ganjoo contributed till his retirement in 2021 in the field of glaciology. He completed twelve research projects funded by premier national funding agencies of India. His field of research was spread to the states of Himachal Pradesh and Jammu and Kashmir (now Union Territories of Jammu and Kashmir and Ladakh). He has published seventy-five research papers in reputed national and international scientific journals including *Quaternary Science Reviews, Climate Change, Palaeo3, Quaternary International, Journal of Asian Earth Science, Asian Perspectives, Zeitschrift fur Geomorphologie, iScience, Global Economics Science*, etc. Besides, he has been part of several research/technical reports and contributed chapters to edited books. He served on several administrative positions in the University of Jammu as Head of the Department of Geology, Founder Director of Institute of Himalayan Glaciology, Director of College Development Council, Registrar and Dean of Planning and Development.

**Professor M. N. Koul** born in 1944, received his post-graduate degree in Geography from the Panjab University, Chandigarh. He worked for his M.Phil. programme with eminent glaciologist Prof. Naseer Ahmad at the University of Kashmir followed by doctoral degree from Panjab University, Chandigarh, on the fluvial and glacial geomorphology of Liddar valley, Kashmir. He was associated with the team from New Hampshire University, USA, led by Dr. Paul Mayweski on glacier expedition to Gangabal (Harmukh, Sindh) and Shafat glaciers (Nun Kun, Suru valley) and earned training in the field of glaciology in his early days. He has seven books and sixty research papers published in refereed national and international journals including *Climate Change*, *American Journal of Climate Change*, *International Journal of Sciences*, *Journal of Ecology and the Natural Environment*, etc., and conference proceedings to his credit. He has supervised nine students for Ph.D. and thirty students for M.Phil. programmes. He has completed seven major research projects in the field of glaciology sponsored by the Department of Science and Technology, New Delhi, and Space Applications Center, Ahmedabad. He monitored six major glaciers of northwest Himalaya, namely Siachen, Durung Drung, Kangriz, Machoi, Kolahoi and Naradu, extending from Ladakh to Himachal Pradesh as the team leader. He remained associated with teaching at the Department of Geography, University of Jammu for 35 years. Prof. Koul has been the President of Indian Institute of Geomorphology (IGI) from 2007 to 2008 and President of Institute of Indian Geographers (IIG) from 2020 to 2023. He is a widely travelled geographer and has attended and chaired various national and international conferences. His major research interests include glacier dynamics, Quaternary glaciology, glacial and fluvial geomorphology, and environment studies.

# Chapter 1
# Introduction

**Abstract** Himalaya, Third Pole, is the largest water tower of Asia that provides 86,000,000 $m^3$ of water annually feeding seven major rivers, namely Ganga, Indus, Brahmaputra, Salween, Mekong, Yangtze and Huang, of the world ensuring round-the-year water to one billion people. The young Himalayan Mountain formed because of the collision of India and Eurasia Plates since ~ 60 Ma after gaining appreciable elevation during Pleistocene became the abode of thousands of glaciers outside the polar region. The variation in the volume, covered area and frontal movement changes due to climate change coupled with various physical determinants of landscape are preserved in the geological records of the glacier valleys. The northwest Himalaya in India comprising Ladakh, Jammu and Kashmir, Himachal Pradesh and Uttarakhand are the focus here.

**Keywords** Indus Molasse · South Tibetan Detachment System · Main Central Thrust · Southwest Indian Monsoon · Agulhas Leakage · *Brokpas* · Silk road · *Dards* · Buddhism

Himalaya, the youngest mountain range in Asia, extends from Afghanistan in north-west to Myanmar in northeast covering a stretch of about 2500 kms. The mountain range is formed because of the collision between India and Eurasia Plate during the Cretaceous period (~ 65 Ma). The break-up of Gondwana landmass led to the disintegration of India Plate from the landmass and subsequently the India Plate travelled approximately 3000 kms to collide with Eurasia Plate (Fig. 1.1a) to give rise to mighty Himalaya. Recent palaeomagnetic and bio-stratigraphic evidence constrains the India-Asia collision in the northwest part of the Himalayan orogen at between approximately 50 and 60 Ma (Leech et al. 2005; An et al. 2021; Bian et al. 2021). The India-Eurasia Plate collision during Cretaceous-Early Tertiary led to the disappearance of Neo-Tethys Sea along two northward dipping subduction zones (Jagoutz et al. 2018). The marine sediments lying in the Tethys Ocean, between India and Eurasia Plates, bore the brunt of squeezing in due to the continuous movement of India Plate over past ~ 60 Ma, resulting in the origin and phased uplift of youngest Himalaya Mountain range of the world. In Ladakh region, Ladakh batholith is separated from



R. K. Ganjoo and M. N. Koul, *The Himalayan Glaciers, Climate Change and Society*,
https://doi.org/10.1007/978-3-031-78283-1_1

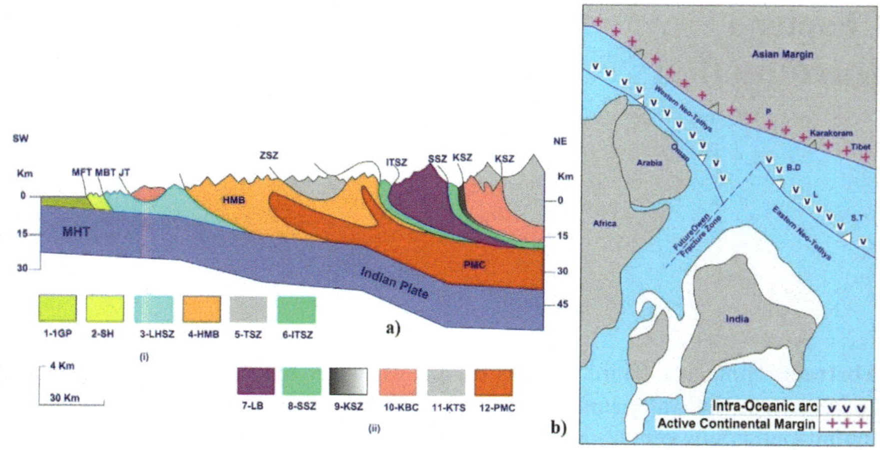


**Fig. 1.1** **a** Geological cross-section through the NW Himalaya and Karakoram. (i) Himalayan Collision tectonic zone: Indo-Gangetic Plains (IGP), Sub-Himalayan Cenozoic foreland basin (SH), Lesser Himalayan Sedimentary Zone (LHSZ), Himalayan Metamorphic Belt (HMB) including Lesser Himalayan Jutogh Nappe (JN), Higher Himalayan Crystalline (HHC) Belt and Tso Morari Crystallines (TMC), Tethyan Sedimentary Zone (TSZ), Indus Tsangpo Suture Zone (ITSZ) and Spongtang Klippe (SK), (ii) Trans-Himalayan tectonic units:-Ladakh Batholith Complex (LB), Shyok Suture Zone (SSZ), Karakoram Shear Zone (KSZ), Karakoram Batholith Complex (KBC), Paleo-Mesozoic Karakoram Tethyan sequence (KTS), Partially Molten Crust (PMC). MFT-Main Frontal Thrust, MBT-Main Boundary Thrust, MCT- Main Central Thrust, ZSZ-Zanskar Shear Zone, MHT-Main Himalayan Thrust (after Jain et al. 2012). **b** Cretaceous paleogeography of the India-Asia convergence zone. B.D. Bangladesh; P. Pakistan (Kandahar, Spin Boldak, Ras Koh); L. Ladakh Himalaya (including Drass, Shergol-Suru valley, Spongtang and Nidar ophiolites) and S.T. South Tibet ophiolites. (after Bhat and Ahmad 2021)

the Karakoram block in northwest by the oceanic sediments and ophiolitic remnants (the Shyok Suture Zone or SSZ). The suture zone rocks in the Ladakh region are Indo-Tibetan Suture Zone (ITSZ), an older suture, comprising Ladakh island arc complex within Neo-Tethys and India, and the SSZ representing a collision zone between Eurasia and an amalgamated India-Ladakh mass to the south dating to Middle-Late Eocene age (~Ma 45–35 Ma) (Tripathy 2011).

Deposition continued regionally along the India-Eurasia collision zone until Late Oligocene-Early Miocene time (~26–23 Ma) (Bhattacharya et al. 2020a; Zhou et al. 2020) consistent with the palaeontological (ostracod) data reported earlier from the correlative sediments near Basgo. Low-temperature thermochronology suggests post-depositional Miocene-Pliocene (~22–4 Ma) cooling of the Indus Basin in central Ladakh. The Indus Basin Sedimentary Rocks (IBSRs)/ Indus Molasse in central Ladakh, excluding the Upper Indus Group, experienced post-depositional cooling from > 170 °C to 200 °C in Miocene-Pliocene time. The ZHe (He detrital zircon) ages suggest that the cooling of basin began between ~ 19 and 8 Ma, whereas thermal modelling proposes the initiation of cooling by ~ 22–20 Ma in the Lower Indus Group of the Indus Basin that continued throughout by ~ 15–12 Ma (Bhattacharya et al.

2020a). Though ZHe temperatures suggest that cooling largely occurred through in early Middle Miocene time, AHe (He detrital apatite) cooling ages supported the cooling continued into the Pliocene time until at least ~ 4 Ma (Bhattacharya et al. 2020b). The cooling may have begun across the Indus suture, earlier in the northern Lower Indus Group Formations between ~ 22 and 20 Ma and then progressed southward in the Tar Group and Lato Formation between ~ 15 and 12 Ma. The erosion by Indus River, draining the India-Asia collision zone in NW India since at least Late Oligocene-Early Miocene time, is identified as a potential driver of the Miocene-Pliocene cooling (Bhattacharya et al. 2020b).

The collision of tectonic plates also resulted into various associated geological activities, such as folding, faulting and volcanism, making the geological and tectonic history of the rising Himalaya complex. The north–south transect of Himalaya Mountain range is tectonically divisible into four major zones, namely trans-Himalaya, Central Crystalline, Lesser Himalaya and Siwalik or sub-Himalayan range. Each tectonic unit preserves the evidence of its geological origin and complexities and is separated from other by thrust, namely South Tibetan Detachment System (STDS) / Trans-Himadri Thrust (THT), Main Central Thrust (MCT), Main Boundary Thrust (MBT) and Main Frontal Fault/ Himalayan Frontal Thrust (MFT/ HFT) (Fig. 1.1b). There is evidence that both structures, the MCT and the STDS, may have been active between ~22 and 19 Ma (Kellett et al. 2018), with MCT and other structures within the Greater Himalayan Sequence (GHS) may have been active more recently (Montemagni et al. 2019). The coarsening upward of Siwalik foreland basin sediments is attributed to the intensification of monsoon, uplift of hinterland because of MCT reactivation, and development of intra-foreland thrust system during ~ 6 and 0.5 Ma (Kumar et al. 2011).

The uplifted Himalaya served as an orographic barrier to wind circulations and the region began to experience southwest Indian monsoons that played decisive role in the landscape evolution and spectacular biodiversity (Manish and Pandit 2018). The Tibetan Plateau massif, also known as the Roof of the World, emerged at an early stage of the collision between India and Eurasia Plates that also played significant role in strengthening the Southwest Indian Summer Monsoon (SWIM) causing intensification of rain and weathering. The elevated Himalaya-Tibetan Plateau and monsoonal rains worked together to reduce the concentration of $CO_2$ in the global atmosphere leading to the global cooling from later Tertiary (~ 10 Ma) till Quaternary (Yasunari 2020) causing the formation of high-altitude glaciers. Changes in global temperature and intensification of monsoon at 10.5, 5.5 and 3 Ma (Sanyal et al. 2010) led to the setting of cyclicity in the glacial and interglacial periods starting from small amplitude of 20,000 to 30,000 years cycle to present 100,000 years cycle (Yasunari 2020). The phenomenon of dry and wet phases is noticed in Ladakh region where the aridity and aeolian activity is recorded during 25–17 ka and < 12–8 ka, and intra-dunal lake formation and fluvial gullying at ~ 12 and 7 ka (Kumar et al. 2017).

## 1.1 The Geology

Ladakh region is bound by Karakoram ranges in the north and Great Himalayan range in the south. The Ladakh range and Zanskar range run parallel to the Karakoram and Great Himalayan ranges and lie in the Ladakh region. The Ladakh and Zanskar ranges are along the Indus River extending from northwest to southeast. Ladakh region experiences cold climate and barren landscape qualifying the region as cold desert. Leh, the capital of Ladakh, at ~ 3500 m amsl reels under extreme annual range of temperature between 34.8 °C (summer) and −27.9 °C (winter) with a mean annual temperature of 7.3 °C (Chevuturi et al. 2018). The high-altitude cold desert region of Ladakh receives sufficient precipitation, in the form of snow, during winter period due to the influence of Westerlies. Besides the precipitation received in the winter period, sparse precipitation during the summer is received from SWIM or ISM that enters the Ladakh valley from south.

Ladakh region lies in the trans-Himalaya tectonic zone and largely preserves the marine sedimentary deposits of Palaeozoic to Mesozoic Era (Fig. 1.2). The granitic intrusions within the sedimentary deposits and meta-sediments are often encountered. The Ladakh batholith, an intermediate rock, is present to the north of Indus River. The Ladakh region preserves the evolution of Ladakh magmatism since 160 Ma into three distinct time periods: i) 160–80 Ma representing intra-oceanic island arc complex dominantly of mafic magmatism, (ii) 80–50 Ma shift in magmatism to felsic composition (involving Indian continental margin) in the form of Kohistan-Ladakh batholith and (iii) 35 Ma to 18 Ma, the occurrence of magmatism in the form of post-collisional dykes (Bhat and Ahmad 2021). Zanskar region comprises two antiforms—Zanskar Crystalline Complex (ZCC) and Tso Morari Crystalline Complex (TCC), with Zanskar Super Group (ZSG) as major synform between the two. ZSG consists of fossiliferous Palaeozoic sandstone and argillaceous rocks, and Mesozoic to Eocene shelf carbonate rocks.

The expanse of Ladakh region between Ladakh batholith in the north and Zanskar range in the south hosted several glaciers during the Quaternary times that have now withdrawn to peaks in the Indus valley. Ladakh region holds the oldest record of Quaternary glaciation in India subcontinent dating back to approximately 430Ka known as Indus valley glacial stage. Geological evidence for other four glacial stages, namely Leh, Kar, Bazgo and Khalling, recorded from the region (Owen et al. 2006), defines the retreat of glaciers from the Indus valley. The fluctuation in glaciers led to the release of meltwater giving rise to the deposits of fluvio-lacustrine sediments (Fig. 1.3a) as thick as ~ 150 m, spread over Ladakh region at punctuations. The liquefaction and/or low intensity earthquakes have been instrumental in developing deformational structures (Fig. 1.3b) in the soft fluvio-lacustrine sediments (Phartiyal and Sharma 2009).

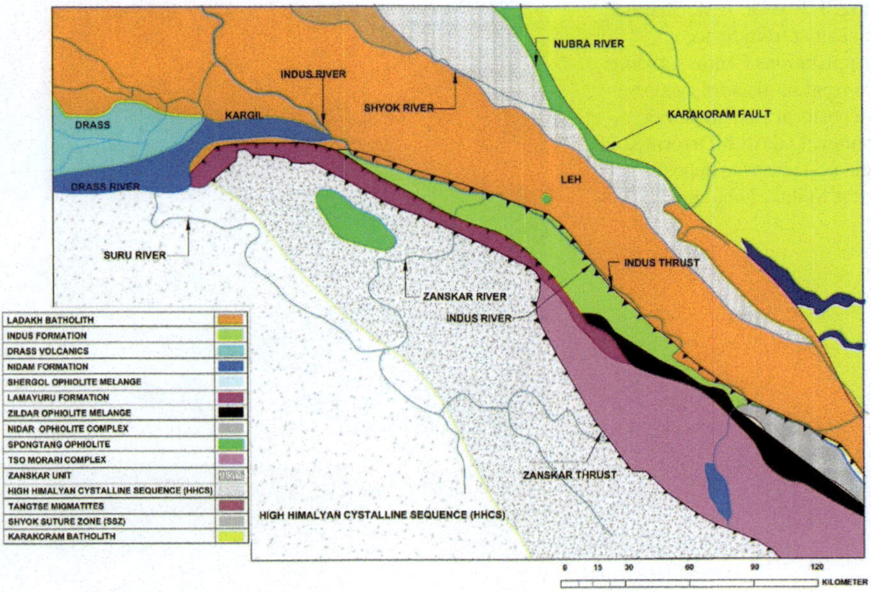


**Fig. 1.2** Geological map of Ladakh Himalaya (after Bhat and Ahmad 2021)

## 1.2 The People and History

Ladakh, the Land of Passes, is also known as *Maryul* or lowland. Fa-Hein called it as *Kia-chha* and Hiuen Tsang called it as *Ma-Lo-Pho* in their works. The *Brokpas* were the first to occupy the lower reaches of Indus valley. Immigrants from Karja (Kulu) settled in Gya and then spread to the area between Martselang and Khaltsi. Dards immigrated to parts of Ladakh from Tibet and Skardu. Dards considered themselves the descendent of Indo-Aryan race. Skit Lde Nemagon, the ruler of Tibet, conquered Ladakh in tenth century AD and established kingdom at Shey till the middle of tenth century AD. The land of Ladakh then was divided into small principalities that were at war with each other. The Ladakh emerged as a strong kingdom under King Singge Namgyal who was known for his statesman and diplomacy. He built historic nine-storey Leh palace and promoted horse polo. King Lalityadita, the ruler of Kashmir, conquered Ladakh in eighth century AD. Ladakh was ruled under a Tibetan regime, Nyima-Gon, from the royal Tibetan house, between eighth and thirteenth centuries that also marked the spread of Buddhism in Ladakh. Footprints of Buddhism reached Ladakh from central India on the way to Tibet. The Buddhist monks, Padmasambhava and Bhikhu Shantarakshita, were the first to build Buddhist monastery in the Suru valley in eighth century AD. The Basgo king, Bhagan, who claimed the title of *Namgyal* (meaning victorious), consolidated the Ladakh empire and formed Namgyal dynasty which existed till recent. The Leh Palace and Hemis monastery were built by the Namgyal ruler. General Zorawar Singh, a Dogra warrior,

**Fig. 1.3** **a** Dr. A.K. Sharma, Scientist from Space Applications Centre standing in front of glacio-fluvial deposits at Khalsi, Nubra valley, Ladakh **b** Deformed glacio-lacustrine deposits near Spituk, Leh, Ladakh

invaded Ladakh in nineteenth century to disrupt the rule of Namgyal. Ladakh acceded to India during the partition and was reorganized into two districts, namely Leh and Kargil.

Ladakh has been a hub for the trade between India, China, Tibet and Central Asia since ancient times. The Upper Indus in Ladakh formed the major centre for traders coming from Kashmir, Punjab and Yarkand. The ancient trade routes moved through the area between Chitral and Uttarakhand enveloping many high-altitude mountain ranges of Pamir, Hindu Kush, Karakoram and the Plateau of Tibet. The caravan of traders using the cross-national trade route played strong role in influencing the cultural, social and economic development of the regions including Ladakh. The cross-national trade route was termed as 'Silk Road' by German geographer Ferdinand von Richthofen in nineteenth century. The early form of Buddhism entered China from India following the trade route. The trade links between India, China and Central Asia formed the backbone of Himalayan economy. The results of genetic investigation of Ladakh population confirms the ethnic and genetic diversity which includes genetic contributions from disparate regions of the continent including West, East, South and Central Asia (Rowold et al. 2016).

Zanskar region is thinly populated with small villages scattered in the region. The largest village, Padum, is the headquarters of Zanskar. Most of the villages are situated along the Zanskar River and its two main tributaries, namely Doda and Lungnak rivers. Earliest evidence of human occupation in Zanskar region, in the form of petroglyphs, goes back to Bronze age. The early occupants are believed to be steppe hunters of Central Asia living between Kazakhstan and China. The *Dards* might have replaced the early settlers. Buddhism from Kashmir spread in Zanskar as early as 200BC as evident from the earliest monuments of Kushan Period (100BC to 500BC). The region was ruled over by Tibetan and imposed Bon religion in seventh century BC. Buddhism regained its foothold between tenth and eleventh centuries and two important monasteries—Karsha and Phugtal—were constructed in Zanskar. The status as independent Buddhist kingdom continued till fifteenth century when Zanskar became subordinate to Ladakh. The invaders from Kulu, Lahaul and Kinnaur destroyed the royal palace at Padum in the year 1822. Subsequently, Zanskar became part of Ladakh region in 1842AD until 2019AD when Ladakh region was declared as Union Territory by the Parliament of India.

Lofty Great Himalayan and Zanskar ranges isolated the Zanskar region, compelling the inhabitants to rely upon themselves and on the scant available resources, until 1979 when road from Kargil connected the region with rest of the world. The inhabitants of inaccessible Zanskar region, in the absence of road, trekked the frozen surface of Zanskar River, known as *Chadar* road, to reach Leh (capital of Ladakh) for trade with outside world. The main occupation of natives of Zanskar is cattle rearing and farming the land they own. The alluvial fans and river terraces form the cultivable land, as most of the terrain is mountainous and cultivable land is scarce. Paucity of natural resources and cultivable land introduced polyandrous marriage practice which served an efficient birth control system to keep the population stable.

Kinnaur is located northeast of Shimla, the capital of the state of Himachal Pradesh of India. The Baspa, Sutlej and Spiti valleys of Kinnaur are surrounded by Zanskar and Himalaya Mountains and Tibet. The Kinnaur Kailash mountain, a natural rock *Shivling*, is the highest peak in Kinnaur. The old Hindustan-Tibet Road passes through Kinnaur valley to enter Tibet at Shipki La pass. It is the second least populous district of Himachal Pradesh, after Lahaul and Spiti.

The early history of the Kinnaur region is obscure and is confined to legends and mythological accounts. Inhabited by tribes, the region was visited by Pandavas of Mahabharat fame. It is believed that the Sutlej River was brought by Kinner King Banasur from Lake Mansarovar to Kinnaur known as *Kinnerdesh* in ancient legends. Kinnaur remained part of the empire of Emperor Kanishka which ended in eighth-nineth century AD with the death of King Harsha. Kinnaur, with Kamru its capital, was part of the erstwhile Bushahr principality which subsequently was shifted to Rampur Bushahr. The Nepalese captured the region in 1810AD after the death of the king of Bushahr. The British troops, subsequently, defeated the Nepalese and region remained under British administration. The principality of Bushahr was dissolved after the independence of India and became part of Himachal Pradesh state.

The Kinnaur region enjoys temperate climate with long winters and short summers. The lower part of Sutlej valley and Baspa valley also receives monsoon with highest rainfall in July month. The upper parts of the valleys, housing glaciers, are in rain-shadow zone and arid. The Baspa River, after rising near the Indo-Tibetan border, flows through Sangla or Baspa valley to meet the major river Sutlej. The last village on the 95-km-long Baspa River valley is Chitkul at an elevation of ~ 3475 m amsl. The Sangla valley, after the name from the village of the same name, is either arable land or pastureland or covered with forest. The peaks surrounding the Sangla valley form the water divide between Sutlej and Yamuna catchments and form border between the states of Uttar Pradesh and Himachal Pradesh of India. Sangla remained a preferred place for traders from Tibet and Garhwal (Himachal Pradesh state) owing to its proximity to both.

## 1.3 Climate Change and Himalayan Glaciers—A Paradox

The geological evidence indicates at least three ice ages in the geological history of planet earth, confirming the repetitious change in climate since inception. The oldest of which occurred at ~ 2.2 billion years ago, in the Pre-Cambrian time followed by widespread glaciations during the Permian time (~ 245–286 Ma), whereas onset of Northern Hemisphere glaciations coincided with global cooling trend between ~ 3.0 and 2.5 Ma, evident in global oxygen isotope records. The West Antarctic Ice Sheet (WAIS) and the coastal margins of the East Antarctic Ice Sheet (EAIS) cooled towards their present polar state, expanding and developing more permanent marine termini and ice shelves which continued through the Mid-Pleistocene Climate Transition (~900–600 kyr) in case of WAIS (Powell et al. 2009).

Most recent ice or glacial age starting at 1.6 Ma (Pleistocene time) comprised glacial periods of the duration approximately 50,000 to 100,000 years separated by briefer interglacial periods of the duration about 10,000 to 50,000 years. The concomitant glacial advances and retreats during Pleistocene time may have been caused by one or more factors, including volcanism, variations in the amount of solar energy reaching certain parts of the earth's surface, meteorite impacts, changes in the global circulation patterns of the earth's oceans and astronomical forcings (Watanabe et al. 2023). Besides, variations in several other drivers, such as North Atlantic Oscillation (NAO), the El Niño Southern Oscillation (ENSO), the Pacific Decadal Oscillation (PDO), the shift in the Inter-tropical Convergence Zone and Indian Ocean Dipole, too influenced the change in climate.

Although changes in ocean circulation pattern, biological productivity, carbon dioxide solubility and other aspects of ocean chemistry have been considered responsible in bringing change in climate, the exact mix of mechanisms is not well understood. The Dansgaard-Oeschger (D-O) events, during the last glacial period, record an abrupt warming of 5–16 °C from cold phase (Greenland Stadial-GS) to a warm phase (Greenland Interstadial—GI) within a few decades (Capron et al. 2021). Synchronizing the Antarctic records to the Greenland ice-core records reveals that GS coincides with the gradual warming phase of the Antarctic Isotope Maxima (AIM) and the GI coincides with AIM cooling phase (Zheng et al. 2021a). The Antarctic trend changes lag the Greenland transitions by about a century (Svensson et al. 2008). The reduced northward transport of heat in Atlantic due to weak Atlantic Meridional Overturning Circulation (AMOC) causes the GS and warming in South Atlantic resulting into AIM warming in Antarctica and giving rise to bipolar seesaw (BPS) (Zheng et al. 2021a). Such variabilities on millennial timescales are indicated as old as 750,000–780,000 years as revealed from the careful study of EPICA's (European Project for Ice Coring in Antarctica) Dome C in eastern Antarctica.

The leakage events of Agulhas Current, based on palaeoceanographic records for past 570,000 years used as proxy, are linked to glacial-interglacial cycles using oxygen isotope stratigraphy, diatom reconstructions and variations in deep-ocean ventilation using carbon isotope ratios and opal fluxes (Civel-Mazens et al. 2021). The Agulhas system (Agulhas Current and Agulhas Return Current) around southern Africa forms a key component of the global ocean circulation that feeds the upper arm of the Atlantic Meridional Overturning Circulation (AMOC) through the leakage of warm and saline waters from the Indian Ocean to the Atlantic (Nirmal et al. 2023). The Agulhas leakage from Indian Ocean to Atlantic is subject to southward shift of Westerlies under warm climate conditions that helps expand the oceanic 'gateway' between the African continent and the STF (Subtropical Front—a zone of active interaction between atmospheric and ocean circulation, which heavily influences the interocean transfer) (Cartagena-Sierra et al. 2021; Nirmal et al. 2023). Strong evidence of Agulhas Leakage Fauna (ALF) for past ~ 450 kyr has been instrumental in the increase at glacial-interglacial transitions of Marine Isotopic Stages (MIS) 12, 10, 8 and 6 (Nirmal et al. 2023). The alterations in the earth's orbital parameters, the position of Westerlies and the amount of solar insolation serve as some of the driving forces in the variability of Agulhas leakage (Verma et al. 2023).

Even though the rise in greenhouse gases due to human activity in past about 150 years is considered the culprit to the increase in global temperature, it is interesting to note that climate changes did not monotonically change during the Holocene rather periodically fluctuated on a centennial scale that might have been associated with solar irradiance (Bae et al. 2022). The most dominant source of energy input for the earth is the Total Solar Irradiance (TSI)—the rate at which energy from the Sun reaches to the earth's atmosphere and varies slightly from day to week—and a little variation in TSI can be responsible for the natural forcing of the earth's climate on global and regional scale (Bhargawa and Singh 2019). Strong positive correlations between solar forcings and terrestrial temperatures started emerging at ~ 4500BP and grew strong during 2000BP–100BP with change in the amplitude of temperatures more in the Northern Hemisphere than that in the Southern Hemisphere (Lu et al. 2022). The average yearly temperature has shown an increase by 0.006691 °C, whereas the trend has accelerated to 0.015502 °C and 0.019795 °C since 1951 and 1971, respectively (Wang et al. 2023). The global dimming-brightening transition shows a decrease over 1961–1981, on global average level, and an increase over

1982–2019 at − 3.07 $Wm^{-2}$ and + 0.33 $Wm^{-2}$ per decade, respectively (Yuan et al. 2021).

Variability in terms of dry/wet records over the Asian monsoon region, such as Northeast China, Eastern China, South Korea and southern Qinghai-Tibet Plateau (Sun et al. 2020 and references therein), is associated with about 500-year cycle of solar activity. Long-term oscillations of the solar background magnetic field are associated with double dynamo waves generated in inner and outer layers of the Sun that is related to the long-term Solar Inertial Motion (SIM)—the motion of the giant planets generates a beat on the Sun in the form of gravity (tidal force) and angular momentum with respect to the solar system's barycenter, called as SIM and total solar irradiance (TSI) (Zharkova et al. 2019). The Solar Activity (SA)—i.e. the emission of heat, electromagnetic waves and particles—is known to change with cycles ranging from several years to decades, centuries and millennia, such as the 11-year Schwabe cycles (also known as sunspot-cycle) and the Gleissberg cycle, which have a characteristic double periodicity of 50–80 years and 90–140 years, respectively (Viaggi 2021).

Since the Industrial Revolution changes in the carbon isotopic composition of atmospheric $CO_2$ due to human activities have caused impact on the natural carbon cycle. Studies suggest decrease in the atmospheric $\Delta^{14}CO_2$ (ratio of $^{14}C/C$ in atmospheric $CO_2$) by ~ 20‰ between 1850 and 1950 (post industrial revolution) as a result of fossil fuel. The intense nuclear weapon testing from the mid-1950s until the mid-1960s rose atmospheric $\Delta^{14}CO_2$ rapidly reaching to highest level of 835‰ in 1964–1965 in the Northern Hemisphere, because most of the bomb tests occurred in the Northern Hemisphere. Decrease in the values of atmospheric $\Delta^{14}CO_2$ is observed since 1990s after reaching the peak in 1964–1965 and is almost linear at about 5‰ $year^{-1}$. Both the $\Delta^{14}CO_2$ trend and the interhemispheric gradient are weaker than expected from fossil fuel emissions alone because of the combined influence on $\Delta^{14}CO_2$ from carbon exchanges with the ocean and land biosphere and by natural $^{14}C$ production and $^{14}C$ emissions from nuclear power plants (Graven et al. 2020 and references herein).

Climate, an intricate dynamic system, is a complex interlink of very large number of degrees of freedom and subsystems that invites challenges for its perception, interpretation and understanding.

## 1.4 Background to the Book

Himalaya, the Third Pole, is source to major and mighty rivers, namely Ganga, Brahmaputra and Indus, in India especially during the dry season. About 800 million people living in the catchment of these mighty rivers survive on the waters released from glaciers of Himalaya (Immerzeel et al. 2010). Besides, the availability of melt-water from the glaciers of Himalaya-Karakoram affects the hydropower generation, agriculture and ecosystem. Around 9500 glaciers, covering an area of over 37,000 $km^2$, in Himalaya-Karakoram region serve as the proxy indicators to measure change in the climate, if any. Monitoring the variable rates of recessions and advances of Himalayan glaciers to record the response of glaciers to climate change is, therefore, a thrust area of research in India (Sood et al. 2020; Singh et al. 2021a).

According to the inventory on Himalayan glaciers prepared by Geological Survey of India, there are altogether 9575 glaciers extending from Ladakh in the west to Sikkim in the east. The northwest Himalaya houses 7240 glaciers located in Indus basin covering an area of 28,191 $km^2$ (Sangewar et al. 2001). The Ladakh (Union Territory of Ladakh) and Kashmir (Union Territory of Jammu and Kashmir) together house 5557 glaciers and more than 93% (5211 glaciers) are confined to Ladakh occupying Karakoram-Zanskar-Ladakh mountains. Himalaya is occupied by a large number of small-sized glaciers, 1–7 km in length, along with some exceptionally large-sized glaciers, e.g. Siachen (73 km long), Gangotri (31 km long), Durung Drung (23 km long) and Zemu (21 km long). Ladakh Himalaya houses 4463 glaciers in two major valleys, namely Indus (2009 glaciers) and Shyok-Nubra (2454 glaciers) (Sangewar and Shukla 2009). Indus, the westernmost Himalayan river, originates from a glacier near Bokhar Chu in Tibet at an altitude of 4164 m amsl in Kailash and flows northwest to enter Leh region. It flows between Ladakh and Zanskar ranges and is joined near Skardu by Shyok River. Himalaya in the northwest extremity is bordered by the Karakoram, Pamir and Hindu Kush mountains.

Recent studies show increase in the winter precipitation of Karakoram-northwest Himalaya (Huang et al. 2022; Narsimha Rao et al. 2021). The winter accumulation in Karakoram Himalaya adds to the health of glaciers that is demonstrated in the gain of mass from 2008 to 2016. Based on ASTER data, glacial mass balance in Karakoram was $-0.06 \pm 0.19$ m w.e.$yr^{-1}$ in 2000–2008 and $0.05 \pm 0.19$ m w.e.$yr^{-1}$ in 2008–2016, whereas the overall glacial mass balance was $-0.03 \pm 0.07$ m w.e.$yr^{-1}$ over the period of 2000–2016 (Lhakpa et al. 2022). Between 2000 and 2012, change in the elevation of glaciers in Jammu and Kashmir (eastern region of the Karakoram) was $-0.19 \pm 0.22$ m $yr^{-1}$, whereas the changes in elevation glaciers for both types, surge- and non-surge, were $-0.18 \pm 0.22$ and $0.20 \pm 0.22$ m $yr^{-1}$, respectively (Vijay and Braun 2018). The large and widely elevated glaciers were stable in the Karakoram Mountains compared to the glaciers in other parts of the Himalaya, as demonstrated during 1986–2018 in the case of Siachen glacier (Kumar et al. 2020), whereas small and high-altitude glaciers showed retreat (Fox-Kemper et al. 2021). The major reason for gain in the mass of Karakoram glaciers is the snowfall that increases with increase in the altitude. Increase in the summer snowfall in the Karakoram protects the glaciers from melting and gaining the mass in the glacier (Farinotti et al. 2020). Decreasing summer temperature and increasing annual precipitation in Karakoram, in last two decades, has been beneficial for glacier ice accumulation, as pointed out earlier in 'Karakoram Glacier Anomaly' hypothesis.

The change in area or shrink rate of glaciers is variable in Indian Himalaya between ~ 0.2 $y^{-1}$ and 0.7%$y^{-1}$ (Immerzeel et al. 2010). The studies of change in area were carried out using multi-temporal remote sensing approach-based study of 121 small and 60 cirque and valley glaciers in Ladakh for the period between 1969 and 2010 revealed that glaciated area of approximately 1000 $km^2$ has decreased by about 14% at the rate of 0.3%$y^{-1}$, in over four decades (Schmidt and Nüsser 2017) with the maximum shrink in 1991 and 2002 at the rate of 0.6%$y^{-1}$, followed by slower rate of 0.2%$y^{-1}$ since 2002 suggesting high decadal variability in the loss of ice cover.

Using high-resolution datasets of Corona and Sentinel-2A shows that glacier area in Alaknanda basin (Central Himalaya, India) reduced to 683 ± 47.81 $km^2$ from 742 ± 44.4 $km^2$ and the number of glaciers increased to 116 from 98 between 1968 and 2020 (Remya et al. 2022) with an annual average recession of 11.75 ± 1.6 $m^{-1}$. On the other hand, smaller glaciers (< 5 $km^2$) in the basin with snouts at lower altitude and high slope show significant loss in area and high rate of retreat. Consequently, significant increase (~38%) in the supraglacial debris cover during 2000–2020 is noted. Besides topographic parameters, increase in winter-time temperature (0.03 °C/year) between 1968 and 2020 is considered the driving factor for deglaciation and fragmentation of Alaknanda basin glaciers.

Estimation of mass balance of 445 glaciers in the upper Alaknanda basin and neighbouring transboundary glaciers using optical stereo imageries from 1973 to 2021 indicates a mean annual area change rate of – 1.14 ± 0.07 $km^2$ $a^{-1}$ and a geodetic glacier mass balance of – 0.34 ± 0.08 m w.e.$a^{-1}$ from 1973 to 2020, leading to an overall mass loss of 12.9 ± 1.7Gt, that accounts for up to 0.036 ± 0.006 mm of sea level rise (Bhattacharya et al. 2023). The increase in summer and annual temperature of ~ 0.6 °C and ~ 0.5 °C, respectively, enhanced the regional glacier mass loss rate from – 0.30 ± 0.07 m w.e. $a^{-1}$ during 1973–2000 to – 0.43 ± 0.06 m w.e. $a^{-1}$ during 2000–2020. However, glacier morphometry, topographic features and uneven debris cover distribution further impacted the regional and glacier-specific mass balance.

Multi-temporal study of glaciers of Jankar Chhu watershed (Lahaul Himalaya, Chandrabhaga basin, India) indicates less rate of recession of 0.17 ± 0.01% $a^{-1}$ from 1971 to 2016 than reported for Chenab basin. The increase in area of debris-covered glaciers between 1971 and 2016 probably due to the melting of clean-ice surface resulting in the exposure of debris cover surface (Das and Sharma 2019). Further, the variability in the rate of recession of glaciers, north facing glaciers receded less than the south facing ones, in Jankar Chhu watershed is a factor of the aspect as northern slopes receive less radiation than the south facing slopes (Patel et al. 2018). Compared to the glaciers of other basins of western Himalaya, namely Chenab, Beas, Miyar, Parbati, Tirungkhad and Baspa, the recession rate of the glaciers of Jankar Chhu is lower for the contemporary period (Das and Sharma 2019).

The debris covering the surface of glaciers serves as non-climate driver to influence the mass balance of the glaciers. Studies in northwest Himalaya confirm that variation of glacier extent in different elevation zones, snout retreat and decadal changes was observed subject to different responses of various glacier types (Bahuguna et al. 2021). The snow avalanches in the high-altitude glaciers of Karakoram and northwest Himalaya, further, interfere with the mass balance of the glacier.

Batbaatar et al. (2018) found an asynchronous behaviour in a cirque glacier of arid Gobi Desert that extended maximum, much after the Pleistocene global retreat in Holocene (~ 8–7 ka) during regional pluvial period. The behaviour is not predicted by the prevailing and generally correct presumption that glacier advances are dominantly driven by temperature, although precipitation also plays a role.

Uncertainties loom large on the projections made on the interrelationship between climate change and change in glaciers in Himalayan region. The book gives a field account of the scientific studies carried out on the glaciers of northwest Himalaya, particularly the glaciers of Ladakh (Union Territory of Ladakh, India), Kashmir (Union Territory of Jammu & Kashmir, India) and Himachal Pradesh (India) to underline the drivers, climatic and non-climatic, that affect the health of glaciers in northwest Himalaya.

This Chapter of the book introduces to the geographical information on part of northwest Himalaya, particularly Ladakh (Union Territory of Ladakh, India), Kashmir (Union Territory of Jammu and Kashmir, India) and Kinnaur (Himachal Pradesh, India). A few representative glaciers from parts of northwest Himalaya housed in Karakoram, Ladakh and Dhauladhar ranges are dealt herein. The landscape punctuated with diverse topography and climate variability forms part of the chapter and also gives an account of the age-old pastoral nomads of the region who sustained the harsh climate conditions to grant a peculiar political, social and cultural identity to the region. The Pleistocene uplift of Himalaya and Tibetan Plateau is considered to have played significant role in setting the climate of Asia vis-à-vis the globe and initiated the formation of glaciers during Quaternary period (Pleistocene and Holocene). Chapter 2 provides a review of the Quaternary glaciations in Himalaya, particularly the northwest Himalaya. The application of Optically Stimulated Luminescence (OSL) and Cosmogenic Radionuclide (CRN) dating techniques, in past three decades, has improved the building of glacial history of northwest Himalaya with up-to-date information on the Quaternary glaciation. Besides, brief account of the geological setting of the Durung Drung, Machoi, Kangriz, Siachen, Kolahai and Naradu glacier valleys in northwest Himalaya with geological age, stratigraphy and lithology is also dealt in the chapter. Climate change in Himalaya is poorly understood because the interplay of tectonics and cryosphere is instrumental in causing variation in the regional climate of high altitudes. Chapter 3 highlights the significance of several factors of climate change, such as temperature, precipitation, diurnal temperature variation, solar irradiation, wind speed, albedo and snow cover, with a special reference to Ladakh (Union Territory of Ladakh), Kashmir (Union Territory of Jammu & Kashmir) and Kinnaur (Himachal Pradesh), that serve as micro-climate conditions to determine the climate of Himalayan region. Further, the effect of the climate variation on the glaciers of the part of northwest Himalayan region is discussed in this chapter. The extent and pattern of glaciation in Himalaya are subject to various natural determinants, such as elevation, aspect, debris cover, hypsometry and slope that are discussed herein. Chapter 4 deals exhaustively on the natural determinants with regard to northwest Himalaya. Glacier valleys of selected basins and sub-basins, such as Liddar, Baspa, Shyok-Nubra and Drass, are dealt herein with long-time and short-time studies of the glaciers in the basins/sub-basins to explain the role of the natural determinants. Also, mass balance study of selected glaciers in northwest Himalaya is discussed with an emphasis on the significance of study to explain the relationship between glacier fluctuation and climate change. Chapter 5 introduces to the geomorphological landforms sculpted by the glaciers and associated agencies in high northwest Himalaya. Undoubtedly, based on geological evidence the cover of snow and ice during Pleistocene period was sufficiently more, in extent and thickness, in Himalaya. The extensive snow/ice cover during Pleistocene period played pivotal role in modifying the landforms intensively to create the present landscape. The variety of landforms created by snow/ice, meltwater and permafrost in the Himalaya are described to give an overview of the northwest Himalayan landscape. It is believed that the frequency and intensity of natural hazards in Himalaya have increased with the change in climate over past few decades causing serious impact on the life and property of mountain people. Chapter 6 deals with several types of natural hazards, such as GLOF, LLOF, cloudbursts, snow avalanches, seismic-induced avalanches and earthquakes, commonly occurring in the high-altitude Himalaya. The chapter introduces to the causes and effects of natural hazards in the backdrop of changing climate in Himalaya. Besides, geological evidence of natural hazards in Himalaya is documented to give geological/ historical account of the disasters. Chapter 7 gives an account of the impact on the livelihood of the mountain people of Himalaya. The chapter gives an insight on the strategies adopted by mountain people of Ladakh, Jammu and Kashmir and Himachal Pradesh on the land use and agriculture practices to cope with the changing climate in the region.

# Chapter 2
# Geological Overview of the Glacier Valleys

**Abstract** The glaciological studies of selected glaciers in Zanskar valley, Suru valley, Baspa valley, Drass valley and Liddar valley are highlighted emphasizing the research input in the field of glaciology for past four decades. The uplift of Tibetan Plateau at ~ 50 Ma and the rapid uplift at ~ 10 Ma resulted in the disappearance of North Pacific Deep Water (NDPW) and shaping of the modern global climate favouring the glaciers of Himalaya. Barring an early report of the oldest glaciation dating radiometrically to late Middle Pleistocene, all other reports of glaciation in Ladakh date back to late Pleistocene and Holocene. The oscillation in late Pleistocene and Holocene climate in Ladakh is concurrent to the changes in Inner Mongolia, northeast India, central Himalaya and central India. Early studies suggested four glacial stages of glaciation in Zanskar valley followed by recent studies suggesting retreat of the snout of Drang Drung glacier by 925 m at the rates 21.11 $my^{-1}$ since 1971 which accelerated to 60 $my^{-1}$ since 2014. Kangriz/Parkachik glacier in Suru valley shows anomalous behaviour with no change in the position of snout but change in the thickness since 1902. Variable rate of retreat is recorded between 1973 and 2018 in the Machoi glacier (Drass valley) with average rate of retreat at ~ 11.43 $my^{-1}$ from 1973 to 1980 which increased to ~ 12.0 $my^{-1}$ from 1980 to 1990, further to ~ 12.3 $my^{-1}$ from 1990 to 2000 and ~ 12.8 $my^{-1}$ from 2000 to 2010. Kolahoi glacier (Liddar valley) has shown retreat from 1857 to 2014 (last 157 years) at average rate of ~ 18.2 $my^{-1}$whereas in another study the retreat from 1962 to 2014 (52 years) is calculated at the rate of 19.5 ± 1.1 $my^{-1}$.

**Keywords** North Pacific Deep Water (NDPW) · Tibetan Plateau · Mid-Pleistocene Transition (MPT) · El Niño · CRN (Cosmogenic Radionuclide) · Last Glacial Maxima (LGM) · Roman Warm Period · Little Ice Age · Durung Drung glacier · Kangriz/Parkachik glacier · Machoi glacier · Kolahoi glacier

Glaciers serve as an effective indicator to understand and measure the impact of global climate change, both in past and in future. Glaciers have been an important source of water to a large population of the world, more so during the periods of reduced precipitation. Monitoring the change in glaciers, therefore, is important



R. K. Ganjoo and M. N. Koul, *The Himalayan Glaciers, Climate Change and Society*,
https://doi.org/10.1007/978-3-031-78283-1_2

than ever before in the present-day scenario of climate change. The scientific understanding, in past three decades, on the factors responsible in driving change in climate has progressed with added scientific information on the cascading effect of climate change on glaciers. The Himalaya, abode of 9575 glaciers, has been an open laboratory for geo-scientists to monitor and assess the changes in the glacier bodies due to the change in climate over a period.

The irregularities/complexities in atmospheric and thermohaline circulation patterns are coupled with the uplift of plateaus and mountains to define significant role in determining the global climate change. The establishment of the monsoon system in East Asia is a result of the uplift of Tibetan Plateau at about 50 Ma that was accelerated at about 10–8 Ma and more recently. The rapid uplift of Tibetan Plateau at 10 Ma played a critical role in the disappearance of North Pacific Deep Water (NDPW) and subsequent shaping of the modern global climate (Ferreira et al. 2018). Some coupled modelling results show that removing of Tibetan Plateau can lead to the collapse of Atlantic thermohaline circulation (Su et al. 2018; Yang and Wen 2020). The numerical experiments shed light on the significance of the absence/presence of the Tibetan Plateau which can lead to as much as 4 °C cooling /warming in surface atmosphere, and about 3 °C cooling/warming in surface ocean (Yang et al. 2020). Even the absence/presence of the Tibetan Plateau can result in about 2 °C warming/cooling in the remote Southern Hemisphere high latitudes. The presence of Tibetan Plateau has weakened the Westerlies in the Northern Hemisphere, increased the poleward atmosphere heat transport and thus weakened the polar vortex. The expansion of Northern Hemisphere ice sheet at the onset and middle of the mid-Pleistocene Transition (MPT), in the context of enhancing global cooling across the MPT (1.25–0.6 Ma), intensified the global Asian climate with the initiation and enhancement of ~ 100 kyr glacial cyclicity (Rohling et al. 2021). Expanded Northern Hemisphere ice sheets started shedding ice-rafted debris into Gulf of Alaska at ~ 1.3 Ma and Central North Atlantic Ocean at ~ 0.9 Ma, respectively, causing decrease in the glacial global average sea surface temperatures markedly (Ao et al. 2023). The thermal forcing of the Tibetan Plateau can also influence remote areas, such as the western Pacific subtropical high and the upper-level jet over the North Pacific (Yang and Wen 2020 and references herein).

More and more scientific investigations explain the variance in the South East Asian Monsoon (SEAM) rainfall particularly due to orogeny (93%) over the last 30 Ma. However, multiple drivers, e.g. precession, obliquity, ice sheet and $CO_2$, impacted SEAM rainfall variability during Plio-Pleistocene (Thomson et al. 2021 and references therein). On the other hand, rainfall variability of Indian Summer Monsoon (ISM) is driven by precession over last 30 Ma. The cave and marine geological records support the dominance of precession on Pleistocene ISM variability (Gebregiorgis et al. 2018). Significant role of orogeny, precession and $CO_2$ in holding a strong control on ISM rainfall is indicated by 39%, 25% and 21%, respectively, (Thomson et al. 2021 and references herein).

The onset of South Asian Monsoon at late Miocene (~10 Ma) in Himalaya resulting in strong erosion and global cooling trend with more seasonal climate at ~ 8 Ma is coincident (Jepson et al. 2021 and references herein) and coupled with internal duplexing (rock uplift) (DeCelles et al. 2020 and references herein). High mountain ranges and plateaus formed as a result of tectonics lead to orographic precipitation (Schildgen et al. 2022) which increases erosion and unroofing of mountain belts resulting in influencing the tectonics (Kaveh-Firouz et al. 2023 and references herein). The studies, however, show decoupling between climate and tectonics (Forte et al. 2022) or weak coupling (Kaveh-Firouz et al. 2023 and references herein).

## 2.1 Ladakh Himalaya

Ladakh is one among the highest inhabited terrain located in the trans-Himalaya. The trans-Himalaya, a tectonic sub-division of Himalaya, lies beyond the main Himalayan ranges towards the north. It is surrounded in the north by three major mountain ranges, Zanskar, Ladakh and Karakoram, and with high Himalaya in the south.

The compositional variation in Ladakh magmatic arc, from mafic towards south to more acidic towards north, is suggestive of northward dipping subduction. The evolution of the Ladakh magmatism from 160 to 80 Ma is dominated by mafic magmatism and represents intra-oceanic island arc (IOIA) complex; from 80 to 50 Ma, the involvement of Indian continental margin in the form of Kohistan-Ladakh batholith is suggested with change in magmatism to felsic composition; and from 35 to 18 Ma, the magmatism occurred in the form of post-collisional dykes (Bhat and Ahmad 2021). The first collision of plates took place along the Shyok Suture Zone (SSZ) between IOIA complex (including Kohistan-Ladakh batholith) and southern margin of Asia during mid-Cretaceous to late Cretaceous. The final collision was during Late Eocene (~40 Ma) along the Indus Suture Zone (ISZ) between the accreted IOIA and Asian plate margin and the northern margin of Indian plate.

The ISZ is the collage of rocks that identifies the northern limit of the Zanskar Zone (ZZ) bound by Zanskar Thrust towards the south and Karakoram strike-slip fault towards the north, whereas SSZ, consisting of Shyok and Nubra volcanics, Shyok ophiolitic mélange, and Saltoro molasse and flysch, separates the Ladakh Batholith-a pre- and post-collisional calc-alkaline magmatism—towards the south from Karakoram Batholith—an Andean-type magmatism towards the north (Sivaprabha et al. 2022; Bhat et al. 2023 and references herein). The linear belt of Karakoram Zone, comprising metamorphic, igneous and sedimentary rock types ranging in age from Permo-Carboniferous to Late Cretaceous, lies north of the Shyok Suture Zone (Bhat et al. 2023 and references herein).

Ladakh falls in the rain-shadow region with annual rainfall merely 80 mm. Two main weather systems affect Ladakh region, namely Western Disturbance (the moisture brought by westerly winds in upper air from Mediterranean, Caspian, Black Sea) in winter and Monsoon (the moisture brought by easterly winds from Bay of Bengal) in summer (July–Sept). Acting as an environmental boundary between North Atlantic and monsoon forcings, the Ladakh region facilitates in deciphering the dynamics of two—Westerlies and Monsoon—significant atmospheric circulation patterns. The studies unfold the impact of Westerlies between 19.6 and 11.1 ka in the region followed by the impact of monsoon forcings till 7.5 ka and regaining of Westerlies from 7.5 to 6.1 ka (mid-Holocene) coinciding with decreasing insolation, weakening monsoon and enhanced El Niño activity (Nag et al. 2023).

The two significant pathways of moisture are important contributors to the hydrological budget of the Ladakh region in the form of precipitation, snowmelt, glacier melt, permafrost and baseflow. The largest cryosphere cover of ~ 21,193 $km^2$ contributes ~ 50% of water to annual river flow of Indus River basin (Azam et al. 2021). Studies suggest that snowmelt contribution is higher in Indus (63 ± 1.2%) and Shyok (58 ± 1.7%) whereas glacier melt contribution is higher in Nubra (64 ± 2.3%) and Suru (60 ± 2.7%) sub-basins/catchments (Lone et al. 2022).

The temperature of the high barren landscape dips as low as −30 °C in Leh and −50 °C in Drass. The freezing low temperature for most part of the year allows sparse vegetation and high-altitude cold aridity. In the short summer season, temperature goes up to 20 °C causing melting of snow. The careful harnessing of the melt water helps to sustain the cultivation. The major river of Ladakh is Indus that originates from near Mt. Kailash and enters from Tibet at Demchok. The major tributaries of river Indus are Zanskar, Yapola, Shyok, Nubra, Hanley and Sangeluma. Some well-known lakes, such as Pangong Tso, Tso Morari, Tso Kar, Yaye Tso, Kiun Tso and Amtitla, are present in Ladakh region.

## 2.2 Nubra-Shyok Valley

The Nubra-Shyok valley houses Siachen glacier, the largest glacier in the Himalaya and second largest outside the polar and sub-polar regions. Siachen glacier and its tributaries cover more than 1500 $km^2$ area and lies between Saltoro range to the west and Karakoram range to the east. The meltwater of Siachen glacier is the main source to the Nubra River which joins the Shyok River near Khalsi to further join the Indus River.

The Nubra-Shyok valley forming a part of the Karakoram terrane (Fig. 2.1a) welded to southern Eurasia during the Early Jurassic and southern margin became magmatically active because of northward subduction of the Meso-Tethyan oceanic lithosphere since Middle Jurassic leading to the development of the Andean-type continental Karakoram Arc (Saktura et al. 2022). Margo volcanics document the extrusive equivalents of Karakoram Arc (batholith) that are magmatic rocks of Mesozoic age, whereas the Nubra-Siachen leucogranites of Miocene age are formed because of the partial melting of Karakoram Batholith (Pundir et al. 2020). The Karakoram Arc and intra-oceanic Shyok ophiolite magmatism were contemporaneous during Jurassic to Cretaceous and the magmatism extinguished with collision in late Cretaceous in both arc systems putting Shyok ophiolite juxtaposed to Karakoram Arc to form Shyok Suture Zone (Saktura et al. 2021). The Nubra-Shyok valley is represented by calc-alkaline metaluminous (I-type) Tirit granites, while the Panamik granites exposed in the Khalsar-Panamik area are weakly peraluminous,

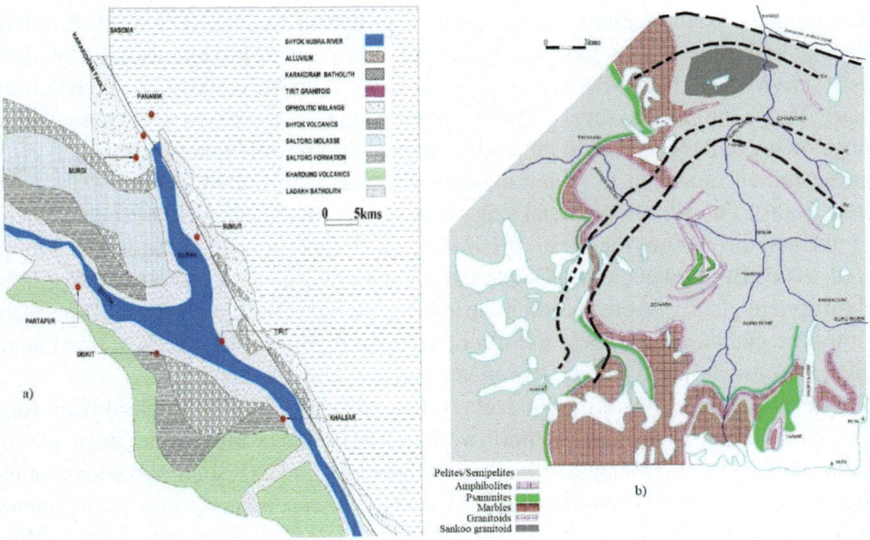


**Fig. 2.1** Geological maps of **a** Nubra-Shyok valley, northern Ladakh (after Bhutani et al. 2009) and **b** part of Suru valley (after Walker et al. 2001) Grt- garnet St- staurolite Ky- kyanite

like those noted for the granites from the Shyok-Murgo (Pundir et al. 2020 and reference herein).

The popular and presently acceptable view is that SSZ is a site of Early to Late Cretaceous Ocean closure and ISZ is a younger main suture between the Indian and Asian continental plates. The interplay of closure of oceans, magmatism and mountain building gave rise to various and divergent interpretations earlier on the SSZ and identified it as an oceanic suture, as a subduction zone that is older than or younger than the ISZ, as a tectonic repetition of the ISZ, as a back-arc complex related to the ISZ and as a marginal basin (Chandra et al. 2023 and references herein).

The evolution of Karakoram, thus, exhibits a geological record stretching from the Precambrian to the Neogene with multiple episodes of crustal shortening, thickening and metamorphism and includes both subduction-related Andean-type magmatism and extensive post-collisional crustal melt S-type granites.

### 2.2.1 *Quaternary Climate and Palaeo-Glaciation of Ladakh-Karakoram Region*

The earliest account of the glacial history of Ladakh range was presented by Fort (1983) and Burbank and Fort (1985) suggesting at least four glacial advances during Quaternary, based on the position of terminal moraines and reconstruction of Equilibrium Line Altitude (ELA). Five stages of glacial advances supported by CRN

(Cosmogenic Radionuclide) dating were identified in Ladakh (Owen et al. 2006) starting with Indus valley glacial stage as the oldest (~ 430 ka) and followed by Leh (~ 200 ka MIS-6), Kar (MIS 5), Bazgo (~ 41 ka to 74 ka MIS-3/MIS-4) and Khalling (early Holocene) glacial stages. The extent of glaciation decreased with every glacial stage owing to the decrease in precipitation due to uplift of Himalaya causing barrier to the Indian monsoon to penetrate northward (Taylor and Mitchell 2000; Owen et al. 2006). The Khalling glacial stage is in sync with the early Holocene advances of glaciers throughout the western Himalaya and Tibet (Owen et al. 2002a; Finkel et al. 2003; Barnard et al. 2004) whereas, on the other hand, high-latitude Northern Hemisphere glaciers were receding during this period (Owen et al. 2006). The early Holocene glacial advance was a result of increase in monsoonal precipitation due to insolation maximum which is identified as a local climate factor.

The mid- to late Quaternary palaeo-climate history, covering a period from Last Glacial Maxima to Recent, of northwest trans-Himalaya region has been reconstructed using biotic (pollens, spores) and abiotic (sediments) climate proxies supplemented with several chronological techniques, e.g. accelerated mass spectrometry (AMS), carbon radiometric dating, optically stimulated luminescence dating (OSL) and cosmogenic radionuclide dating (CRN). Blöthe et al. (2014) suggested at least two older episodes of massive infilling in the region during late Pleistocene at ~ 530 ka followed by pronounced incision at ~ 200 ka, and younger episodes between ~ 50 and 20 ka. Proxy evidence suggests increase in moisture conditions between 12.9 and 12.5 ka followed by extremely weak moisture conditions at ca. 12.2–11.8 ka and again peak monsoon between ~ 10.9 and 9.2 ka (Demske et al. 2009). High resolution isotopic investigations corroborate to increase in ISM precipitation between ~ 13.1 cal.ka and 11.7 cal. ka and highest during the ~ 11.2 cal.ka–8.5 cal.ka (Mishra et al. 2015).

Shukla et al. (2020) suggested three glacial advances of decreasing magnitude in the Puche valley (near Leh), southern Ladakh ranges with Puche glacier Advance-1 (PGA-1) the oldest and radiometrically dated to the early Marine Isotopic Stage (MIS)-2 (31.5 $\pm$ 3.3 ka) that persisted until around the global Last Glacial Maxima (LGM) (22.7 $\pm$ 1.6 ka), the PGA-2 dated to the post LGM (14.4 $\pm$ 1.0 ka) and undated PGA-3 to the youngest (? Holocene age). At variance with the previous suggestions that recommended monsoon-driven glacier advances, the present studies suggest that Westerlies played role in driving the glacier advances during the MIS-2 in the southern Ladakh ranges.

The scientific study of glacial lakes suggests cold conditions from 5400 cal.yr BP to 4800 cal.yr BP with severely cold and dry conditions between 4800 cal.yr BP and 4400 cal.yr BP in Ladakh Himalaya (Phartiyal et al. 2020). Lei et al. (2014) based on biome reconstruction, and Wünnemann et al. (2010), based on decline in the detrital input of lakes, advocated for the reduction in the availability of moisture and reduced meltwater influx, respectively, after ~ 7000 cal.yr BP. Decrease in monsoon dynamics and setting in of dry and cool conditions since ~ 4700 cal.yr BP are recorded from lakes (Morrill et al. 2006; Demske et al. 2009) and between 4400 cal.yr BP and 3900 cal.yr BP in Qinghai lake (Shen et al. 2005). The colder climate between 4300 cal.yr BP and 3800 cal.yr BP, a regional cool phase in Ladakh, was instrumental to commence neo-glaciation activity (Phartiyal et al. 2020). The 4300 cal.yr BP coincides with the Holocene Climate Optimum (Wu and Liu 2004) and maximum aridity in Holocene (Fontes et al. 1993). Slightly warm condition is witnessed from 4400 to 3400 cal.yr BP punctuated with cold events at 4200, 3800, 3600 and 3000 cal.yr. BP (Phartiyal et al. 2020). The 4.2 ka is termed as an event of weak monsoon based on the studies at Tso Kar lake, Ladakh (Wünnemann et al. 2010), Daihai lake, Inner Mongolia China (Xiao et al. 2019), Mawmluh Cave speleothem, northeast India (Berkelhammer et al. 2012), Lonar Lake, central India (Anoop et al. 2013, Menzel et al. 2014; Prasad et al. 2014), Sahiya Cave, North India (Kathayat et al. 2017), Lake Rara, Nepal Himalaya (Yasuda and Tabata 1988; Nakamura et al. 2016) and Central Higher Himalaya (Phadtare 2000). The 4.2 ka global dry event terminated with distinct wet climate conditions (Cullen et al. 2000; Weiss and Bradley 2001; Staubwasser et al. 2003). The proxy evidence suggests amelioration in climate conditions between ~ 4134 and 3720 cal.yr BP (Phartiyal et al. 2020). Between ~ 3000 and 2644 cal yr BP, once again, cold climate conditions prevail followed by a warm period from ~ 2633–2166 cal.yr BP in Karakoram Himalaya (Phartiyal et al. 2021) which indicates dominant summer monsoon (Thompson et al. 1997; Ji et al. 2005).

The lake studies from Qinghai (Ji et al. 2005; Liu et al. 2006, 2008) and western China suggest rise in lake level after ~ 2400–1700 cal yr BP (Liu et al. 2016) corresponding to the 'Roman Warm Period' (150 BC to AD 270). In Ladakh, warm and oxidizing climate conditions are recorded from 2064 to 1182 cal.yr BP and the warm phase continues till 958 cal.yr BP followed by cold climate from 958 cal.yr BP to 644 cal.yr BP, an expression of the 'Little Ice Age' over trans-Himalaya (Phartiyal et al. 2020). The continued warm conditions till 419 cal.yr BP led to the increase in aridity (reduction of precipitation) and warming (increase in temperatures) causing loss of snow/ice cover over entire Ladakh range (Phartiyal et al. 2020).

The monitoring of snout of Siachen glacier in Nubra valley, the largest glacier in Upper Indus Basin (UIB), has provided a fluctuating record of the movement of snout for over a century (Raina and Sangewar 2008). The glacier advanced by about 725 m between 1862 and 1909 AD at an average rate of 15.42 $my^{-1}$ followed by retreat of about 50 m between 1909 and 1929 AD at an average rate of 2.50 $my^{-1}$. The glacier, once again, experienced retreat of about 400 m between 1929 and 1958 AD at an average rate of 14 $my^{-1}$ and continued to be stable till 1985AD. Negligible retreat at an average rate of 3 $my^{-1}$ was experienced by the glacier between 1985 and 2004 AD (Fig. 2.2). The average rate of retreat further reduced to about 0.6 $my^{-1}$ (Ganjoo and Koul 2009). The ISM penetrates UIB infrequently as it is situated at the boundary between tropical and continental climate influence but is recipient of precipitation in winters from the Westerlies originating from Mediterranean and Caspian Sea (Fowler and Archer 2005).

Yet another study to analyse change in area of 121 small-sized glaciers and to measure the retreat of 60 cirque and valley glaciers in the trans-Himalayan Kang Yatze Massif, Ladakh, northwest India, between 1969 and 2010 provided a scenario of high variability in the change of glaciers (Schmidt and Nüsser 2012). The studies reveal that, over the last four decades, the glaciated area decreased by about 14%

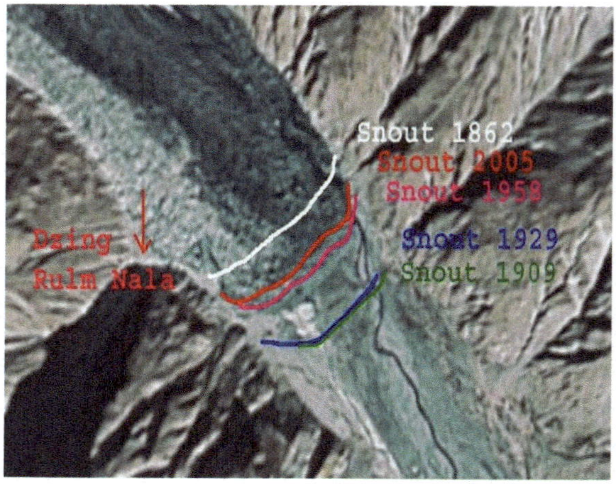


**Fig. 2.2** Relative approximate positions of snout of the Siachen glacier between 1862 and 2005

($0.3\%y^{-1}$) from 96.4 to 82.6 $km^2$, the average retreat of ice front amounts to 125 m (3 $my^{-1}$) and ice cover loss shows a high decadal variability with decrease rate of 0.3% $y^{-1}$ between 1969 and 1991, with a maximum shrink of 0.6% $y^{-1}$ between 1991 and 2002 followed by $0.2\%y^{-1}$ since 2002. In the glaciers of size smaller than 0.75 $km^2$, 26 lost more than 30% and 14 lost less than 4% of their initial area. The study further reveals that glaciers, on an average, lost 8% of their length and one-third of the glaciers, belonging to all classes of size, retreated less than 5%. The oldest advance of glacier in Nubra-Shyok valley is dated to $60.4 \pm 5.2$ ka and is termed as Tirith-II (MIS-4) followed by a stage of pulsating decrease in ice volume at around $42.0 \pm 3.0$ ka (mid-MIS-3) (Ganju et al. 2018). The second advance, termed as Tirith-I, at $30 \pm 2.5$ ka (beginning of MIS-2) continued till $18.2 \pm 1.8$ ka (Last Glacial Maximum) followed by deglaciation after $15.8 \pm 1.6$ ka until the early Holocene ($10.3 \pm 1.3$ ka). The third minor glacier advance, termed as Siachen Glacial Advance (SGA), took place in mid-Holocene ($6.8 \pm 1.0/7.2 \pm 1.4$ ka), facilitating the Siachen glacier to extend below the snout, and the youngest glacier advance, dated between $1.0 \pm 0.4$ ka and $0.5 \pm 0.2$ ka, allowed the glacier to expand a few hundred meters below the present-day snout (Ganju et al. 2018). The OSL dates from glacial deposits at 30–90 km below the Siachen glacier in Nubra valley suggest extensive glaciation of the valley between 24 and 18 ka, a period corresponding to the LGM (Nagar et al. 2013). The recent observations suggest that the Siachen glacier does not show any alarming retreat after the Little Ice Age (Ganju et al. 2018) which is in close conformity with earlier studies (Ganjoo and Koul 2009; Nagar et al. 2013). However, the glacier advances in Nubra-Shyok valley are attributed to decrease in temperature and increase in moisture contribution due to the enhanced mid-latitude Westerlies (Nagar et al. 2013; Ganju et al. 2018).

The comparative study of 114 glaciers of Nubra valley on the basis of Survey of India (SOI) topographic sheets of 1969 AD, Landsat imageries of 1989 AD and IRS LiSS III imageries of 2001 AD suggests that the glaciers have not changed much in their length and area between 1989 and 2001 AD compared to the time period between 1969 and 1989 AD (Ganjoo et al. 2014). The study for a period between 1969 and 1989 AD suggests considerable decline in the area from 994.99 $km^2$ in 1969 AD to 932.90 $km^2$ in 1989 AD, whereas negligible decline in the area from 932.90 $km^2$ in 1989 to 930.37 $km^2$ in 2001 is recorded (Ganjoo and Koul 2013). Accordingly, Siachen glacier records change in the maximum length from 68.30 km in 1969 to 67.56 km in 1989 to 67.50 km in 2001 (Ganjoo and Koul 2009). Interestingly, 25 glaciers have gained the area, 26 glaciers have vacated the area and 33 glaciers do not show any change between 1989 and 2001AD in Nubra valley (Ganjoo et al. 2014). Three glaciers show gain in area between 1969 and 2001 AD whereas eight glaciers show gain in area from 1969 to 1989AD followed by recession from 1989 to 2001 AD. On the other hand, five glaciers receded between 1969 and 1989 AD but show advance between 1989 and 2001 AD. The changes in the glaciers of Nubra valley, therefore, are varied, complex and not necessarily due to climate change.

Further studies of glaciers in Himalaya and Karakoram for the period 2001–2010/11 revealed that 1752 (86.8%) glaciers are stable in position, 248 (12.3%) glaciers show retreat in the snout positions whereas 18 (0.9%) glaciers have advanced (Bahuguna et al. 2014) confirming the asynchronous behaviour of glaciers with respect to latitude, elevation and climate/weather variation from region to region.

## 2.3 Zanskar-Suru Valley

Zanskar region covering an area of approximately 7000 $km^2$ is situated at an altitude between 3500 and 7000 m amsl. The region is drained by Zanskar River which is fed by two main tributaries—Doda and Lungnak rivers. The Doda River originates near Pensi La Pass (4400 m amsl) and flows towards Padum, the headquarters of Zanskar. Two main tributaries, namely Kurgiakh-chu and Tsarap-chu, join to form Lungnak River. The Zanskar River further joins the Indus River near village Nimmu short off Leh. Great Himalayan range separates Zanskar region from Kishtwar and Chamba regions, and Zanskar range separates it from Ladakh region. The mountain ranges surrounding Zanskar region protect it from most of the monsoons resulting into high-altitude semi-arid desert. The solid precipitation, in the form of snow, falls during the harsh and long-lasting winter season of 7–8 months.

The Zanskar and Spiti regions of Tethyan Himalaya comprises Phanerozoic sequence of rocks ranging in age from late Precambrian to lower Eocene and overlies the Central Crystalline rocks of High Himalaya. Mesozoic Zanskar supergroup represents the passive continental margin which is separated from Palaeozoic Lahaul supergroup by a rift of continental theollites and mildly alkaline flood basalts of Permian age. The Mesozoic Neo-Tethys passive margin is represented by carbonate platforms of Triassic age and fossiliferous shales of Jurassic age. The Triassic carbonates deciphering shallow marine conditions continued to deposit till late Cretaceous followed by Kangi-La flysch and fossiliferous limestone of Eocene age forming the youngest shelf deposit of Zanskar range.

The major episode of alkaline magmatism all along the Tethys Himalayan Zone was followed by deposition of widespread glauco-phosphoritic horizons during a period of transgressive pulses associated with depleted oxygen in the world ocean (Oceanic Anoxic Events OAE, 1c and OAE 2). The 65 Ma- to 70 Ma-year-long sedimentary cycle of the Tethys Himalaya passive records first a deepening and then a shallowing-upward trend, with a repetition of sedimentary facies controlled by a complex interplay of phenomena—final fragmentation of Gondwanaland, widespread magmatic activity, increased $CO_2$ content in the atmosphere, global tendency to eustatic rise, 'greenhouse' climates, latitudinal shifts due to continental drift, modified patterns of palaeoceanographic circulation, anoxic oceanic events in the world oceans and related effects on the biosphere—active at both global and regional scales (Silva et al. 1992 and reference herein). The Zanskar Shear Zone (= South Tibetan Detachment) in the northern margin of the Zanskar crystalline separates the high-grade Haimanta of the Zanskar crystalline from low-grade Haimanta and overlying Tethyan rocks including rift-related Permian Panjal volcanics (Di Pietro 2004 and reference herein).

The Suru valley lies between Pensi La Pass, the entrance to Zanskar valley, and Sankhoo village (Fig. 2.1b) of which the Nun Kun Massif (Nun-7135 m amsl and Kun-7087 m amsl) is a part. The base of the Palaeozoic-Mesozoic sequence of the Zanskar Tethys Himalaya, represented by low-grade meta-sedimentaries and granites, is exposed in Suru valley. Although the Spiti-Zanskar basin is apparently 300 to 400 kms distant from Kashmir basin, the litho- and bio-facies of the basins from Cambrian to Carboniferous period are similar because Kashmir basin was in the strike extension of the Spiti-Zanskar basin (Bhargava 2015).

The closure of Neo-Tethys Ocean followed by collision of Afro-Arabian-Indian plate towards south with Eurasian plate towards north led the evolution of Alpine-Himalayan mountain system (Dilek and Furnes 2009) and subsequent Tethyan ophiolite belt (Ghosh et al. 2017). The dismembered ophiolite sequences (Nidar ophiolite, Spongtang ophiolite, Shergol ophiolitic slice, Suru valley ophiolitic slice and Dras-Thasgam ophiolitic slice) comprising basal residual mantle peridotites, ultramafic–mafic cumulates, isotropic gabbros, dolerite dykes and basic volcanics interlayered with sedimentary rocks occur as discrete slices along the ISZ from southeast to northwest (Bhat et al. 2021 and references herein) and correspond to Early Jurassic to Mid-Cretaceous age (Buckman et al. 2018 and references herein). The Shergol and Suru valley ophiolitic mantle peridotites preserve portion of ancient Neo-Tethys Ocean lithospheric mantle that was subsequently trapped in a subduction zone environment, thereby evolved by an episode of metasomatism before emplacement at present positions (Bhat et al. 2019).

### 2.3.1 *Quaternary Climate and Palaeo-Glaciation of Zanskar-Suru Valley*

The Zanskar basin is studded with 11 major glaciers, namely Durung Drung, Haptal, Haskira, Kange, Hagshu, Sumcha, Lechan, Denya, Mulang, Yaranchu and Gompi covering an area of 1098 km$^2$ (Table 2.1). The summer snowline (Equilibrium Line Altitude—ELA) rises transversely from SW to NE but is slightly lower on NE than on SW side. Accordingly, the ELA is above 5800 m amsl on the south but at 5500 m amsl on the north. The present-day climate of Zanskar valley is arid to semi-arid. The Durung Drung glacier is situated on Zanskar fault South Tibetan Detachment System (Dezes 1999) that separates the Himalayan Crystalline Zone from the meta-sediments of Tethyan Himalaya. The reactivation of Zanskar fault in the younger geological times, late Pleistocene-Holocene, has been responsible for the block movement on the west (Pensi La Pass) compelling the glacier to turn north and further northeast (Ganjoo 2009) (Fig. 2.3). This has resulted in a moderate dip of 25° in the east part of the glacier body. The glacier body shows overlapping of layers in the west caused due to the tectonics in the area giving a false impression of surge in glacier. Durung Drung or Drang Drung glacier, the second largest after Siachen glacier (Negi 2002), flows southwest of Pensi La Pass—the Gateway to Zanskar.

Osmaston (1994), on the basis of geomorphic evidence, identified four stages of glaciations in Zanskar (Zangskar) valley with Kilima glacial stage (M1) as the oldest one, followed by Tongde glacial stage (M2), Tepuk glacial stage (M3) and Durung Drung glacial stage (M4). Further, based on the remnants of conspicuous moraines, belt of hummocks just above Rangdum Gompa and steep moraine ridge near Sankhoo village, palaeo-glaciation in Suru valley is recognized (Osmaston 1994) which is not correlatable with Quaternary glaciation of Zanskar valley in absence of radiometric dates. Taylor and Mitchell (2000, 2002), adopting the names of glacial stages of Lahaul valley Owen et al. (2001), argued for extensive glaciation during Chandra glacial stage in Zanskar whereas the glaciers were confined to tributary valleys during the subsequent Batal and Kulti glacial stages. Eight glacial advances, namely Skio I, Skio II, Chaluk, Hankar, Nimaling I, Nimaling II, Gapo Ri I/Dzo Jongo I/Kang Yaze and the Gapo Ri II/Dzo Jongo II/Tasken Ri II, in order of antiquity from oldest to youngest, were recognized in Zanskar range (Damm 2006) with no numerical dating.

Three glacial stages, namely Lato (~244–49 ka), Shiyul (25–15 ka) and Kyambu (3.4–0.2 ka), dating back to at least ~ 244 ka are recognized from Lato Massif, Zanskar range (Orr et al. 2018). The glaciation with each stage has changed from expanded ice caps to piedmont glacier and finally to cirque and valley glaciers with a retreat of ~ 16 km upstream and shift of ELA from ~ 5050 to 5650 m amsl over the past 244 ka. The reduction in precipitation reaching interiors of Himalaya is considered effecting the maintenance of positive glacier mass balance (Finkel et al. 2003; Bookhagen et al. 2005; Owen et al. 2008; Seong et al. 2009) emphasizing the role of regional climate controls in bringing change in glacial cycles in many of the semi-arid regions of the Himalayan–Tibetan orogen (Orr et al. 2018).

**Table 2.1** Morphometric parameters of Suru-Zanskar valley glaciers

| Name of the glacier | Length (in kms) | Breadth (mean) (in kms) | Area (in $km^2$) | Snout direction | Snout position (in m amsl) | Relative relief highest (in m) | Relative relief lowest (in m) |
|---|---|---|---|---|---|---|---|
| Durung Drung | 23.75 | 4.43 | 136.43 | NE | 4182 | 6050 | 5116 |
| Haskira | 15.0 | 0.56 | 22.18 | NE | 4250 | 5600 | 4925 |
| Kange | 17.75 | 3.26 | 34.34 | NE | 3980 | 5680 | 4830 |
| Hagshu | 15.02 | 1.96 | 75.00 | N | 3900 | 5680 | 4790 |
| Sumcha | 10.5 | 2.36 | 12.87 | NE | 4475 | 5800 | 5137 |
| Lechan | 6.00 | 1.26 | 4.65 | N | 4510 | 5960 | 5235 |
| Denyai | 5.3 | 1.06 | 6.50 | NE | 4750 | 5800 | 5275 |
| Mulung | 6.5 | 4.1 | 26.73 | NE | 4180 | 5400 | 4790 |
| Haptal | 13.00 | 5.2 | 61.48 | NE | 4575 | 6010 | 5292 |
| Yaranchu | 9.25 | 3.7 | 25.94 | NE | 4180 | 6020 | 5100 |
| Tingzumche | 4.00 | 3.76 | 16.91 | NW | 4800 | 5480 | 5140 |
| Gompi | 5.75 | 0.16 | 10.61 | NE | 4780 | 5520 | 5150 |
| Chabar | 2.75 | 1.33 | 1.90 | NE | 4920 | 5600 | 5260 |
| Shimiling | 6.2 | 1.3 | 6.4 | NE | 4740 | 5920 | 5310 |
| Pholoklow | 4.5 | 4.00 | 6.54 | NW | 4760 | 5280 | 5020 |
| Chogo | 9.75 | 1.66 | 14.58 | NE | 4550 | 5600 | 5525 |
| Khapang | 7.5 | 1.00 | 4.74 | NE | 4700 | 5820 | 5260 |
| Koromshe | 7.4 | 1.33 | 6.74 | NE | 4760 | 5880 | 5320 |
| Temasi | 2.5 | 1.00 | 2.63 | NW | 4600 | 4920 | 4760 |
| Churalp | 6.00 | 4.5 | 5.50 | NW | 4640 | 5520 | 4640 |
| Tidu | 6.5 | 8.3 | 18.67 | E | 4600 | 5689 | 4872 |
| Nateo | 11.00 | 5.16 | 22.21 | NE | 4670 | 5440 | 5055 |
| Katkar | 10.5 | 8.00 | 26.09 | NE | 4520 | 5950 | 5235 |
| Riie | 5.2 | 2.5 | 3.98 | N | 4230 | 5720 | 4975 |
| Tingdur | 4.00 | 4.6 | 8.32 | N | 4660 | 5460 | 5060 |
| Giambal | 10.5 | 5.66 | 23.69 | NE | 4680 | 5920 | 5300 |

Based on the geomorphic evidence, such as moraines, tills and recessional moraines Sharma et al. (2016) proposed SZS-4 (Southern Zanskar Glacial Stage) stage, driven by a combination of enhanced mid-latitude Westerlies during the MIS-4 and/or strengthened ISM during the earlier part of pluvial MIS-3, as the oldest for both Suru and Zanskar valleys. According to them, the SZS-3 stage probably began around the LGM (mid-MIS-2) and continued till the later part of the MIS-2 that represents a period of intensified mid-latitude Westerlies corresponding to

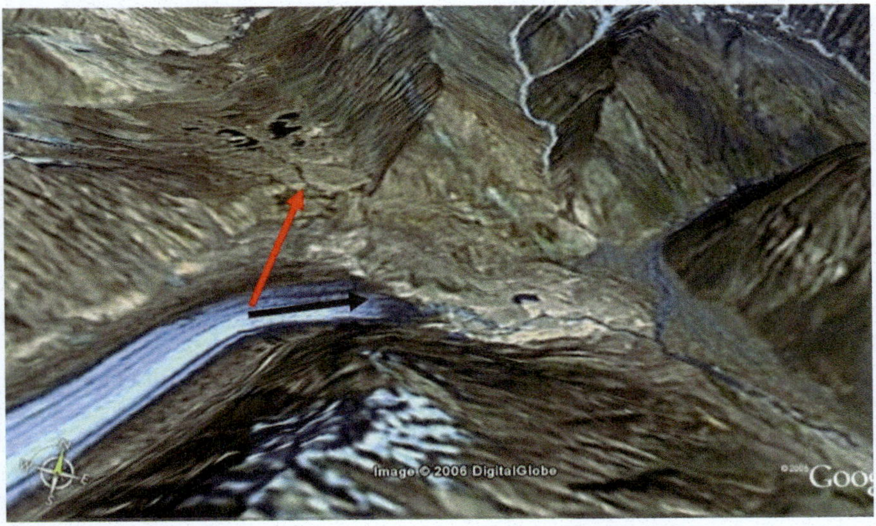


**Fig. 2.3** Shift in the path of Durung Drung glacier (red arrow = palaeo-path, black arrow = present path) during Quaternary period

LGM (Sharma et al. 2016) or later part of MIS-2 during which a gradual strengthening of ISM (Indian Summer Monsoon) is suggested (Schulz et al. 1998). The SZS-2 stage, known for moderate ISM and cooler air temperature, is speculated to have occurred during the early to mid-Holocene period (Sati et al. 2014). It is also suggested that pulsating decrease in ice volume manifested by the presence of around eight curvilinear recessional moraine ridges with the youngest dated to ~ 5 ka led to the eastward deflection of the Durung Drung glacier snout towards the Doda valley during the deposition of SZS-2 moraines. Further, the youngest SZS-1 stage is dated to LIA (Little Ice Age). The outwash gravels and loess–palaeosol deposits dated to c. 1.5 ka, c. 3.4 ka and c. 5.2 ka, supported by the dating of drumlins and lake deposits to ~4 ka and ~3 ka (Sharma et al. 2016) represent warm, dry and humid intervals (Lee et al. 2014) that caused deglaciation under warm climate conditions.

The multiproxy analysis ($\delta^{13}C$, organic carbon, TOC, Rb/Sr ratio, $\delta^{18}O$, LOI) of sediments of Pensi La Pass, Zanskar valley, suggests drier climatic conditions at ~ 6200–4500 cal.yr BP followed by improvement in climate during ~4500–3400 cal.yr BP that ameliorated further between ~3400 and 1500 cal yr BP (Ali et al. 2020). The overall gradual ameliorating period is punctuated by three prominent dry phases at ~3200, 2500 cal yr BP and 1750 cal.yr BP. The mid-Holocene (~6200 cal yr BP–4500 cal.yr BP) cool and dry phase has been attributed to an overall weakened ISM (Sharma and Shukla 2018) causing the youngest millennial-scale cooling in Zanskar (SZS- 1 glacial stage). During the last ~1500 cal.yr BP, a gradual improvement in climatic conditions is seen with a dry spell at ~400 cal yr BP (LIA) (Ali et al. 2020).

Durung Drung glacier experienced a total recession of 311 m between 1975 and 2008, at a rate of ~ 9 $my^{-1}$ (Byrne 2009). There was an increase in debris cover at the tongue of the glacier by > 10% between 1990 and 2006 with the rate of increase by 0.7%$y^{-1}$ between 1990 and 1999, and 0.6%$y^{-1}$ between 1999 and 2006. The increase in debris-covered area of Durung Drung glacier is attributed to the decrease of clean-ice areas in the transition zone between clean and debris-covered ice which is indicative of a receding glacier (Bolch et al. 2008). The study of secular movement of terminus of Durung Drung glacier between 2005 and 2009AD reveals insignificant change in the snout position that is sufficiently corroborated by field evidence and study of satellite imageries (Fig. 2.4) (Ganjoo and Koul 2013). Rashid and Majeed (2018) indicated that Drang Drung glacier shrunk by 13.84% from 1971 to 2017 and snout of the glacier has retreated by 925 m since 1971 at the rate of 21.11 $my^{-1}$ but the retreat accelerated since 2014 at the rate of 60 $my^{-1}$.

The Dalung glacier (DG) present in Suru River catchment has shrunk by $2.6 \pm 0.56$ $km^2$ ($19.2 \pm 4.1$%) at a rate of $0.05 \pm 0.01$ $km^2y^{-1}$ whereas the Padam glacier (PG) present in Tsarap River catchment of Zanskar basin has shrunk by $3.4 \pm 0.65$ $km^2$ ($12.7 \pm 2.4$%) at a rate of $0.06 \pm 0.01$ $km^2y^{-1}$ (Mir et al. 2018a). Accordingly, recession of the front was recorded as $1.240 \pm 0.02$ km ($15.7 \pm 2.0$%) at a rate of 20 $\pm$ 0.4 $my^{-1}$ for DG and $2.690 \pm 0.5$ m ($21.3 \pm 2.1$%) at a rate of $50.7 \pm 0.1$ $my^{-1}$ for PG.

Parkachik or Kangriz glacier in Suru valley, fed by the peaks of Nun Kun Massif, flows northward and abruptly debouche in the Suru River at an elevation of ~ 3560 m amsl. The glacier is located in a large icefield from which different glaciers, such as Shafat, Techafarka, Branktan, Z and Pinnacle, fan out. The accumulation zone of

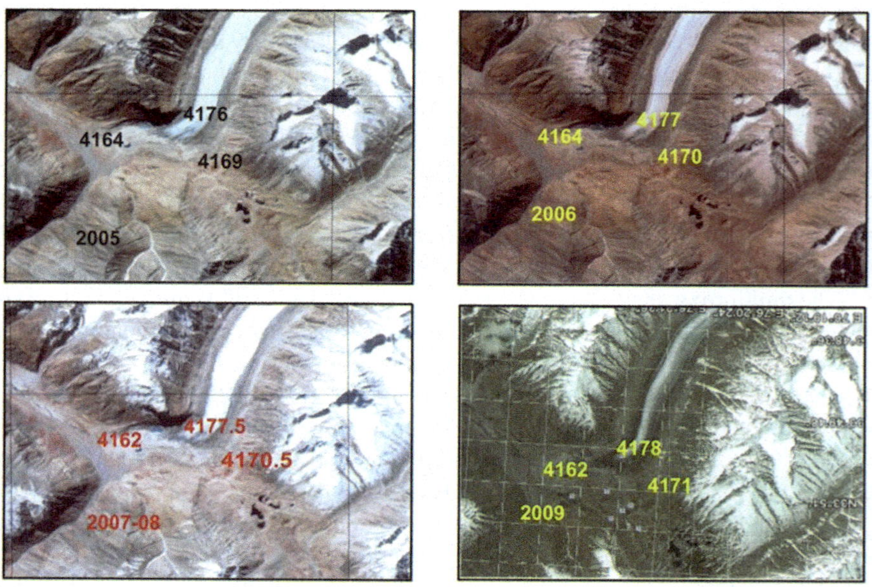


**Fig. 2.4** Change in the position of snout of Durung Drung glacier from 2005 to 2009 studied from satellite imageries (numbers indicate elevation in m amsl)

the glacier is situated in the High Himalayan Crystalline that forms a continuation of the Zanskar ranges. The 13.48-km-long Kangriz glacier emanates from the Kangriz-Shafat icefield and covers an area of 39.94 $km^2$.

The Suru basin was having ~173 m ice thickness and 178 ± 67$km^3$ ice volume during the late Quaternary glacial history of the basin since MIS-3 (dated 33 ka), MIS-2 or LGM (dated 23 ka), late glacial and Younger Dryas (dated 16 ka to 11 ka), early and late Holocene (dated 10 ka–7.3 ka and 2.8 ka) until the LIA (dated 0.7 ka–0.4 ka) (Kumar et al. 2021a).

The Kangriz/Parkachik glacier, like that of some glaciers in Karakoram with the highest levels of relief and which cover extreme altitudinal ranges (Hewitt 2011), shows anomalous behaviour. The tendency of some glaciers of Karakoram, during the 1990s and early 2000s, was to advance sporadically that is also noticed in Kangriz glacier which underwent a large retreat during the 1980s and large advance in the following decade. The catchment of Kangriz glacier, relatively a short glacier, lies at ~ 6000 m amsl with one of the highest accumulation zone in Ladakh region. The two factors, highest accumulation zone and short steep nature of glacier, are considered to account for the sudden and aberrant behaviour of Kangriz glacier because the factors allow to collect snow which passes glaciers on lesser peaks and the shorter lag time of a short steep glacier results in increased level of precipitation (Byrne 2009).

In a detail study on Kangriz /Parkachik glacier, Byrne (2009) recorded a retreat of ~ 141 m between 1979 and 1990, at an average rate of ~ 13 $my^{-1}$. The glacier advanced 157 m between 1990 and 1999, at an average rate of 17 $my^{-1}$ and continued till 2004 at an average rate of ~ 15 $my^{-1}$. The glacier witnessed a negligible retreat of 7 m between 2004 and 2008 at an average rate of ~ 1.75 $my^{-1}$. Interestingly, the debris cover on Kangriz glacier increased by < 8% during the period of advancement between 1990 and 2004 at a rate of ~ 0.6%$y^{-1}$. The Kangriz glacier does not show any change in the position of snout since 1902AD (Ganjoo and Koul 2013 and references herein). However, it is evident from the comparison of the field photographs of 1902AD and 2007AD that the thickness of glacier at snout and ablation zone has certainly reduced in past one century with no retreat in the position of the snout, meaning that the Kangriz glacier has undergone some change in its volume but not in the position of snout (Fig. 2.5) (Ganjoo and Koul 2013). Recent multiparametric study (area, terminal retreat, debris cover, snow line altitude, surface dynamics) of Kangriz glacier for the period from 1971 to 2018 has revealed an overall shrink of 3.3 ± 1.6%, with an expansion in the supraglacial debris cover by 45% (Garg et al. 2022a). They further identified regional warming factors, such as increase in maximum temperature by 7%, increase in minimum temperature by 43% and decrease in precipitation by 3%, effecting the degeneration of glacier. In another study (Mir and Majeed 2018), the Parkachik (Kangriz) glacier has retreated by 127 ± 0.09 m, i.e. (0.75 ± 0.07%) at a rate of 2.9 ± 0.004 $my^{-1}$ with a simultaneous decrease in area from 49.5 $km^2$ to 48.8 $km^2$, i.e. 740 ± 0.7$m^2$ (1.5 ± 0.09%) at a rate of 74 ± 0.7 $m^2y^{-1}$ from 1971 to 2015. The rate of glacier recession of 3.9 ± 0.004 $my^{-1}$ with a corresponding area loss of 500 ± 0.74 $m^2$ (1 ± 0.1%) during 1999 and 2015 was higher than the retreat rate of 2.3 ± 0.001 $my^{-1}$ and an area loss of 240 ± 0.02 $m^2$ (0.48 ± 0.08%) during 1971–1999.

**Fig. 2.5** Documentation of snout of Kangriz glacier in 1902 and 2007. Suru river is in the foreground

The long-term monitoring of 150 glaciers of Drass sub-basin of Suru valley, from 1965 to 2013 by Koul et al. (2016), shows decrease in the area from 187.9 $km^2$ in 1965 to 158.42 $km^2$ in 2001 and further to 156.65 $km^2$ in 2013. An area of 28.48 $km^2$ has been vacated by the glaciers between 1965 and 2001 in comparison with 1.77 $km^2$ between 2001 and 2013. An area of 0.136 $km^2y^{-1}$ was vacated between 2001 and 2013 with 80% of glaciers showing no change in area and in stable mode. The studies further show that snout of the glacier facing N and NW did not show any change in area between 1969 and 2001.

The monitoring of snout of Machoi glacier from 1973 to 2018 showed variable rates of retreat of the snout for 45 years (Taloor et al. 2021). The snout retreated by 80 m at an average rate of ~ 11.43 $my^{-1}$ from 1973 to 1980 which increased to ~ 120 m between 1980 and 1990 with an average rate of ~ 12.0 $my^{-1}$. The retreat increased further to 123 m at an average rate of ~ 12.3 $my^{-1}$ from 1990 to 2000 and to 128 m from 2000 to 2010 at an average rate of ~ 12.8 $my^{-1}$. The retreat of Machoi glacier snout decreased to 112 m at an average rate of ~ 14 $my^{-1}$ between 2010 and 2018. The total retreat of the snout of Machoi glacier was 563 m in 45 years (1973 to 2018) at an average rate of 12.51 $my^{-1}$. The studies (Taloor et al. 2021) also indicated total reduction in the area of Machoi glacier by 0.16 $\pm$ 0.008 $km^2$ from 1973 to 1980. The reduction in glacier area, once again, shows a variable rate of 0.15 $\pm$ 0.0034 $km^2$ from 1980 to 1990, 0.08 $\pm$ 0.0034 $km^2$ from 1990 to 2000, 0.18 $\pm$ 0.0034 $km^2$ from 2000 to 2010 and 0.59 $\pm$ 0.0005 $km^2$ from 2010 to 2018.

## 2.4 Baspa Valley

Baspa River valley lies on the eastern side of Satluj valley in the Higher Himalaya exposing Haimanta and Vaikrita group of rocks intruded by Rakcham granite referred to as Kailash-Kinnaur granite of Ordovician age. The Eocambrian-Middle Cambrian Haimanta Group comprises grey phyllites, quartzite, greenish shale and carbonaceous slate, whereas early Middle Proterozoic Vaikrita Group consists of sillimanite-bearing schist, quartzite, gneiss and migmatite (Dutta et al. 2018). Baspa River flows through lower Palaeozoic rocks and again through Precambrian meta-sediments

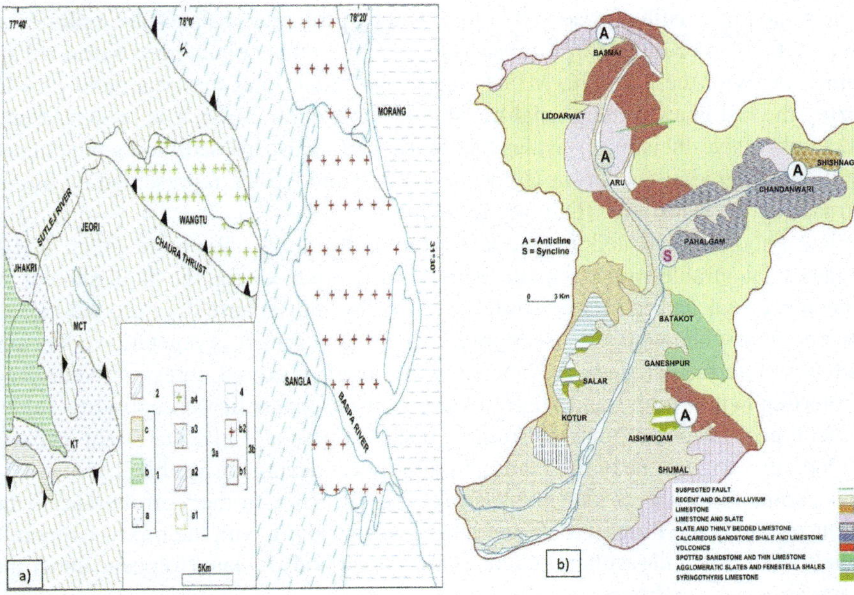


**Fig. 2.6** Geological maps of **a** Sutlej-Baspa valley, Himachal Pradesh. 1: Lesser Himalayan (LH) Rampur Group: (a) Manikaran Quartzite, (b) Rampur Volcanics, (c) Carbonaceous phyllite. 2: Kulu Bajura Nappe augen mylonite. 3: HHC: (a) Jeori Group: staurolite/garnetiferous schist, banded biotite gneiss, augen gneiss (a1), amphibolite (a2), quartz mica schist (a3) and Wangtu granite gneiss/granite (a4), (b) Karcham Group: garnet/staurolite/kyanite/sillimanite schist/gneiss, calc silicate, augen gneiss and migmatite (b1), Akpa leucogranite (b2). 4: Tibetan Sedimentary Zone (TSZ). KT: Kulu Thrust, MCT: Main Central Thrust, VT: Vaikrita Thrust. R: Rampur, K: Karchham (after Jain et al. 2000) and **b** Liddar valley, Kashmir (after Kaul 1990)

intruded by Kinnaur Kailash (Fig. 2.6a). The Kinnaur Kailash granite is separated from high-grade paragneiss and migmatites by Sangla Detachment equivalent to South Tibetan Detachment System(Burchfield et al. 1992).The Baspa River, further, transects Karcham Normal Fault and part of Proterozoic granite gneiss of the Larji-Kullu-Rampur Window near the confluence with Sutlej (Hager 2003, Vannay et al. 2004).

### *2.4.1 Quaternary Climate and Palaeo-Glaciation of Baspa Valley*

The Baspa glacier with a length of 18 km and a surface of ~ 37 km$^2$ is the largest in the Beas and Sutlej valleys of Himachal Pradesh. Late Pleistocene-Holocene depositional environment is accounted for Sangla or Baspa valley by identifying two phases of aggradation initiated well before 49 ka and terminated around 45 ka (49.0 ± 3 ka and 45.0 ± 8 ka) in the form of development of a large alluvial fan at Sangla town area in the pre-existing broad valley followed by prolonged palaeo-lake sedimentation (Ganjoo and Koul 2005, Koul and Ganjoo 2009) that commenced before 23 ka, coinciding with the MIS-3 to MIS-1 transition (28 ka), and ceased around 11.4 ka during the last glacial-interglacial transition (Dutta et al. 2018). The breach of lake led to excessive incision of avalanche debris and lacustrine sediments punctuated by fluvial aggradations during Holocene at c. 10 and 6.5 ka, represented by fluvial terraces by the trunk river (Dutta et al. 2018).

The palynological and magnetic susceptibility study (Ranhotra et al. 2018) suggests favourable moist climate conditions at 16.6 ka for the flourishment of broadleaved oak and conifer woods followed by less moist phase since around 13.3 ka corroborated with magnetic susceptibility values favouring less vegetation cover and increase of surface erosion. Advancement of broadleaved oak forest, good amount of pteridophytes (ferns) and slight lowering in the magnetic susceptibility values depict high monsoon regime interrupting the arid phase at ~ 11.5 ka. Slight increase of Quercus with decline of the Pinus and fern spores during the age bracket of ~ 8.7–7.8 ka coincides with ~ 8.2 ka global cool event. The study further corroborates to the colonization by the Pinus, Abies and other conifers along with the mixed temperate broadleaved taxa and resultant retreat of the glacial body by late Holocene under the averaged warm conditions.

Another study, from a tributary valley to the Baspa River, unravels palaeo-climatic history between 10,450 ± 310 and 1800 ± 64uncal.yr BP (Chakraborty et al. 2006). Relatively warm and moist conditions are interpreted between 10,450 ± 310uncal.yr BP and 4510 ± 56 uncal.yr BP, whereas sediments between 4510 ± 56uncal.yr BP and 1800 ± 64 uncal yr BP were devoid of pollen, but rich in charcoal and sediments younger than 1800 ± 64 uncal.yr BP were rich in steppe vegetation (Chakraborty et al. 2006). The sediments in lowest part of the palaeo-lake, resting underneath the Baspa Hydel Project at the bridge to Chansu, provide radiocarbon date of 7586 ± 99 cal.yr BP, whereas the uppermost layers provide a date of 5072 ± 205 cal.yr BP representing the final stage of lake filling (Draganits et al. 2014). The lower part of palaeo-lake is characterized by the abundance and diversity of angiosperm trees and shrubs, particularly evergreen Fagaceae, and ferns and fern allies, all of which indicate moist and temperate conditions, and the upper part apparently reflects a drier environment or an alpine steppe vegetation with abundant grassy and herbaceous vegetation (Draganits et al. 2014).

## 2.5 Liddar Valley (Kashmir)

Liddar valley (Kashmir) lies in the Upper Indus Basin of northwest Himalaya and spreads over an area of about 1260 km$^2$. The valley ranges in elevation from 1500 to 5500 m amsl and houses about 56 glaciers, including the Kolahoi glacier, the largest glacier (~11 km$^2$) in the Jhelum Basin. Liddar River, the main tributary of Jhelum basin, emanates from the Kolahoi glacier. The valley exhibits a typical temperate climate and receives precipitation in the form of both rain and snow. Eight glaciers, referred as Kolahoi group of glaciers, occupy the west Liddar valley near Kolahoi mountain and group of fourteen glaciers, as the Sheshram group of glaciers, occupy the east Liddar valley along the precipitous mountain slopes of Kunyirhayan, Dudhal and Katsal. Kolahoi glacier, considered a single glacier, has two parts draining into two watersheds (Liddar and Sindh) (Mir 2018) whereas Rudolph Glacier Inventory indicates five parts of the glacier (Rashid et al. 2020).

Kolahoi glacier is about 5 km long and it flows from the northwest of a *neve* field covering about 35 $km^2$. The *neve* field also feeds three other glaciers, the easternmost of these flows northward and its melt water enters the Sind River whereas the other two terminate in the hanging valley glaciers along the eastern wall of Kolahoi glacier. The Kolahoi valley is semi-circular in outline with its head having northwest aspect and terminus having southeast at Pahalgam. The glacier flows over the shales and limestones, overlying Muth Quartzite, of the southern limb of the Basmai Anticline (Fig. 2.6b). Further down Basmai stream joins the West Liddar River and flows through Panjal Traps. The subglacial and lateral moraines cover almost the entire glaciated part of the valley. The primary snout of the glacier faces north whereas the secondary snout is towards east. The uplift of Panjal Traps, due to tectonics over last 4 Ma led to blockade of southwest monsoon winds and precipitation, changing the climatic conditions of the Kashmir valley (Basavaiah et al. 2010).

### 2.5.1 *Quaternary Climate and Palaeo-Glaciation of Liddar Valley*

Ten tributary glaciers joined the Kolahoi glacier during Pleistocene: four joined between the glacier snout and the Basmai stream junction, other four joined below Aru and two terminated as hanging valley glaciers. The geomorphological evidence indicating glaciation of the Kolahoi glacier valley includes the glacial till at Pahalgam and Aru, and terminal and lateral moraines at Liddarwat, Satlanjan and Kolahoi Gunj valley. The climate conditions of Kashmir valley fluctuated between cold arid and warm semi-arid during late Quaternary (Ahmad and Chandra 2018).

Geomorphological and comparative documentary records provide a historical account of retreat of Kolahoi glacier by > 1609 m from 1857 to 1909, 800 m from 1912 to 1961, and 210 m from 1961 to 1984 (Kaul 1990 and references herein) (Fig. 2.7). Geomorphic evidence from Kolahoi glacier valley suggests that the glacier covered an area of 650 $km^2$ during Pleistocene which is now reduced to about 2.5 $km^2$, and net decrease of glacier area is 2.88 $km^2$ in past 42 years, from 13.57 $km^2$ in 1963 to 10.69 $km^2$ in 2005 (Kanth et al. 2011). Recent studies (Rashid et al. 2017) suggest retreat of Kolahoi glacier by 2.85 km during the last 157 years from 1857 to 2014 with an average of about 18.2 $my^{-1}$, and the shrink was 2.81 $km^2$ during the last 51 years (1962–2013) losing an ice volume of 0.30 $km^3$. Shukla et al. (2017) calculated the retreat of Kolahoi glacier, from 1962 to 2014 (52 years), by 1014 $\pm$ 64 m (19.5 $\pm$ 1.1 $my^{-1}$). Accordingly, the change in the area in past 52 years has been from 14.46 ± 0.34 $km^2$ in 1962 to 11.28 ± 0.21 $km^2$ in 2014, amounting an overall deglaciation of 3.18 ± 0.34 $km^2$ or 22.99 ± 2.3%. Further, disparity in the rate of retreat of primary snout (19.5 ± 1.1 $my^{-1}$) and secondary snout (11.34 ± 1.1 $my^{-1}$) is also noticed (Shukla et al. 2017). The studies further reveal variable rate of recession of the glacier for the period from 1962 to 1980 (267.7 ± 62 m i.e. 14.9 ± 3.5 $my^{-1}$), from 1980 to 2000 (356.2 ± 30 m i.e. 17.8 ± 1.5 $my^{-1}$) and from 2000 to 2014 (390.1 ± 1.8 m, i.e. 27.8 ± 1.8 $my^{-1}$), propounding gradual increase in the rate of retreat in past 52 years. The studies by Lone et al. (2018) suggested reduction in the Kolahoi glacier area from 11.75 $km^2$ in 2001 to 10.40 $km^2$ in 2010, thereby confirming deglaciation of the Kolahoi glacier valley.

The studies suggested that two parts of Kolahoi glacier further fragmented into four parts between 1990 and 2000 due to the deglaciation (Rashid et al. 2020). Studies also suggest that retreat rate of the glacier from 0.26%$y^{-1}$ for the period 1962 to 2000 tripled to 0.76%$y^{-1}$ for the period 2000 to 2018. The rate of retreat is variable for

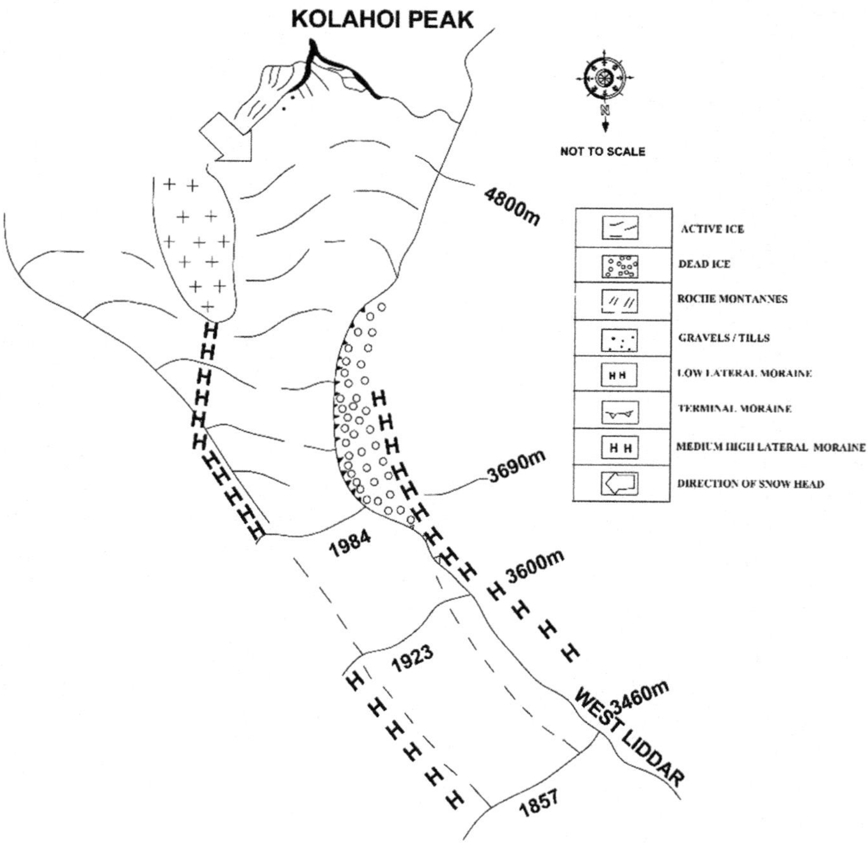


**Fig. 2.7** Sketch showing change in the position of the snout of Kolahoi glacier between 1857 and 1984 (after Kaul 1990)

both the different periods and different parts of the glacier. The rate of retreat of two fragments of glacier is 19.5 $\pm$ 1.4my$^{-1}$ and 19.1 $\pm$ 1.4 my$^{-1}$, respectively, between 1962 and 1992. The rate of retreat further decelerates between 1992 and 2000 to 6.1 $\pm$ 5.3 my$^{-1}$ and 2.8 $\pm$ 5.3 my$^{-1}$, respectively, whereas between 2000 and 2013 the rate of retreat picks up again to 22.6 $\pm$ 3.2 my$^{-1}$ and 20.3 $\pm$ 3.2 my$^{-1}$, respectively (Rashid et al. 2020). The retreat of snout of two different fragments of Kolahoi glacier is highest between 2013 and 2018 at the rate of 24.4 $\pm$ 6 my$^{-1}$ and 27.1 $\pm$ 6 my$^{-1}$, respectively. Significant increase in the temperature (Tmin and Tmax) and small decrease in the precipitation caused the reduction in total seasonal snowfall over the area which may have attributed to the increase in the rate of ablation and less accumulation making a profound negative impact on the glacier health in the region that is manifested in the observed changes in the glacier (Murtaza and Romshoo 2017).

## 2.6 Summary

The Quaternary glacial histories of Lahaul and Garhwal (Barnard 2004), Hindu Kush (Owen et al. 2002b), Nanga Parbat (Phillips et al. 2000), Khumbu Himal (Finkel et al. 2003) and Tibet (Thompson et al. 2006) are asynchronous with the Quaternary glacial histories of Northern Hemisphere glaciers (Scherler et al. 2010) suggesting that the glaciers of Himalaya were influenced by south Asian monsoon (Owen et al. 2002a, Finkel et al. 2003). Maximum advance of glaciers in the Himalaya occurred early in the last glacial cycle, ~ 60–53 kyr (MIS-3), with relatively warm climate and increased monsoon precipitation (Owen et al. 2002a). Glaciers in many parts of the Himalaya are advancing and also growing thick, a positive change, that is ascribed to the enhanced monsoon precipitation associated with global warming (Owen 2009; Ganjoo 2021). The interplay of ISM and Westerlies /Western Disturbances during Quaternary period, therefore, has been governing the change in climate in north-west and west Himalayan region triggering local geological processes, the glacial process being the dominant, to sculpt the present landscape of the region (Sharma and Phartiyal 2018; Ganjoo 2021).

# Chapter 3
# Climate and It's Variables

**Abstract** The interplay between Indian Summer Monsoon, Winter Monsoon and Westerlies determines the glaciation in Himalaya, but local meteorological factors and topographic variations in the mountainous region serve as an effective tool in spatio-temporal variation in the glaciers. Overall, interrelationship between atmospheric circulations at Himalaya, Western Ghats and Tibetan Plateau and oceanic circulations at Indian Ocean, Arabian Sea and Bay of Bengal are instrumental in shaping the climate of the subcontinent. The western Himalaya, particularly Ladakh, Zanskar and Karakoram, has been influenced in past by the Western Disturbance and changes in the driving tool in future shall significantly impress the glaciation of Karakoram-Hindu Kush Himalaya region. The rise in the intensity of precipitation associated with Western Disturbances in Karakoram Mountain by 10.17% with high percentage of snowfall amounting to 64.5% in recent decades is a testimony to it. The large variation in temperature in High Himalaya and trans-Himalaya is due to the local topography of the sites that dictates different thermodynamical and orographic forcings at various locations in the mountain. As such, differential rates of warming are noticed across Himalaya, 0.6 °C/decade over Nepal between 1977 and 2002, 0.2 °C/decade over entire northeast India between 1971 and 2005, 0.16 °C/decade over north-western Himalayas, 0.23–0.43 °C/decade over western Himalayas and 0.27–0.54 °C/decade over Pakistan Himalayas.

**Keywords** North Atlantic Oscillation · Indian Ocean Dipole · ENSO (El Niño Southern Oscillation) · SPI (Serial Peripheral Interface) · Slope Environment Lapse rate—SEL · Inter-Tropical Convergence Zone (ITCZ) · Pir Panjal range · Dhauladhar range

Weather is the physical condition or state of the atmosphere at any point of time. As those conditions change, so does the weather. Climate is a composite of weather conditions over a considerable time. Principal elements of weather and climate are (a) temperature (b) pressure and wind, and (c) moisture and precipitation. Temperature is the intensity of degree of heat. The heat energy is the result of insolation or radiation from the sun. Radiant energy from the sun, which is commonly called heat and light,



R. K. Ganjoo and M. N. Koul, *The Himalayan Glaciers, Climate Change and Society*,
https://doi.org/10.1007/978-3-031-78283-1_3

is transmitted mainly from short waves, whereas long wave or infrared radiation leaves the earth. Solar energy serves as engine that drives the winds and ocean currents, generates the weather and makes the earth a livable place for biological life. The atmosphere is heated primarily from the surface by the absorption of terrestrial infrared radiation and by turbulent heat transfer. There is a great variety and localism of climate in highlands, particularly in mountainous areas related to altitude and exposure changes.

Northwest Himalaya has been called the radiation window of the earth, since thin dry air at higher altitude facilitates the entry of solar radiation by day and rapid loss of heat by terrestrial radiation at night. The rapid day-time heating and night-time cooling results in large diurnal ranges of temperature that are characteristics of the highland climate. At night, during the long winter season (November to May), the terrestrial radiation is rapid and leads to a cooler environment along the ridge slope that induces temperature and precipitation gradient in the winter season favouring sustenance of the glaciers. The steep temperature gradient during summer season forces the cool environment along the ridges that helps sustain the glaciers. The degree of slope steepness varies in the three longitudinal sub-divisions of the Himalaya-eastern Himalaya, central Himalaya and western Himalaya—following temperature slope environmental lapse rate from east to west that exhibits regional characteristics of the sub-divisions. Local winds occur because of diurnally varying thermal differences setting up local pressure gradient (mountain valley winds) or due to topographic barrier on airflow crossing it. The location of Himalaya in the subtropical zone acts as an effective barrier to monsoonal winds as well as to Westerlies.

## 3.1 Glacier Valley Orogeny

The Himalayan ranges are the result of geological events of the collision between the Lesser Himalayan, in the south, and Tethys Himalayan, in the north, tectogens dating back to ~ 60 Ma. The Himalaya attained its present elevation at ~ 15 Ma that led to the initiation and establishment of the monsoon system in Asia. The pulsating tectonic uplift of Himalaya continued till the Holocene period leading to the activation of Lesser Himalaya that greatly influenced the climatic characteristic of the region. The Himalayan Mountain and the offshoots are divided, based on altitude, into three macro-regions represented by three major fold axes. They are the higher Himalaya above 5400 m amsl known as cryospheric cold desert and housing fourteen highest peaks of the world including some of the largest glaciers. The Great Himalaya above 3000 m amsl encompasses snow and ice cirques and lastly the Lesser Himalaya macro-region that lies above 1500 m amsl and extends almost uninterruptedly throughout its length.

Mountain functions as an effective meteorological tool as well as topographic barrier. The snow on Himalaya turns into compact ice mass after compression and following the ice-forming processes through years to ultimately result in the formation of glaciers. Glaciation is strongly controlled by Southwest Monsoon (Indian

Summer Monsoon), Northeast Monsoon (Winter Monsoon) and cold Westerlies. The Himalayan region is prone to severe weather during winter due to large amount of snowfall produced by Western Disturbances (WDs)/Westerlies, and the interplay of WDs with the topography determines the spatial and vertical distribution of precipitation in the western Himalaya (Nischal et al. 2022).

## 3.2 Climate and Glacier Valley Relationship

The tropical glaciers of Himalaya sustain because of the orography. The steep mountain slopes, in the catchment, favour cooler environment along the ridge slope that induces temperature and precipitation gradient in the winter season. The steep temperature gradient during summer season further corroborates to the cool environment along the ridges that supports sustenance of the glaciers. The occurrence of Himalaya in the subtropical zone acts as an effective barrier to both the monsoonal winds and Westerlies (Fig. 3.1a). The funnel-shaped physiographic alignment of eastern Himalaya in the vicinity of Cherrapunji ranges leads to the convergence of maritime winds arising from Bay of Bengal and Arabian sea. The monsoonal winds, thus triggered, result into vigorous and highest rainfall of 1000 cm in Assam, Arunachal Pradesh and adjoining areas in Brahmaputra valley. The maritime air mass then traverses the monsoonal wind to Ganga and Punjab plains. The high monsoon rainfall is characteristic of tropical monsoon climate due to the physiographic alignment of ranges and high relief topography along with close basin morphology that helps monsoon tropical winds to influence as an effective meteorological barrier (Fig. 3.1b, c).

The alignment of western Himalaya Mountain, confined to rain-shadow zone of Zanskar, Ladakh and Karakoram mountains, guides Westerlies air mass arising from Caspian Sea to traverse through Pamir Pass and enter high altitude passes of Karakoram, Zanskar and other adjoining areas before turning along the axis of northwest and central Himalaya. The Himalaya Mountain elevations ensure that no large-scale atmospheric circulation, either monsoon or Western Disturbances, reaches above 600 hpa (hectopascals), also called millibars level, without orographic uplift under normal circumstances. The regional mountain climate and local climate of the glacier are unequivocally influenced by the orographic processes and the changes occurring to the glacier system impact the hydrology and ecosystem process of the lower nival monsoon regime too. Change in the regional climate with respect to the change in Indian Summer Monsoon has a unique effect on the subcontinent. The interrelationship between atmospheric circulations at Himalaya, Western Ghats and Tibetan Plateau, and oceanic circulations at Indian Ocean, Arabian Sea and Bay of Bengal are instrumental in shaping the regional scale climate. The interaction between atmosphere, ocean, land surface, cryosphere and biosphere on various time scales, vis-a-vis global natural phenomena such as El Niño Southern Oscillation North Atlantic Oscillation and Indian Ocean Dipole, effects the climate change (Mayewski et al. 2020).

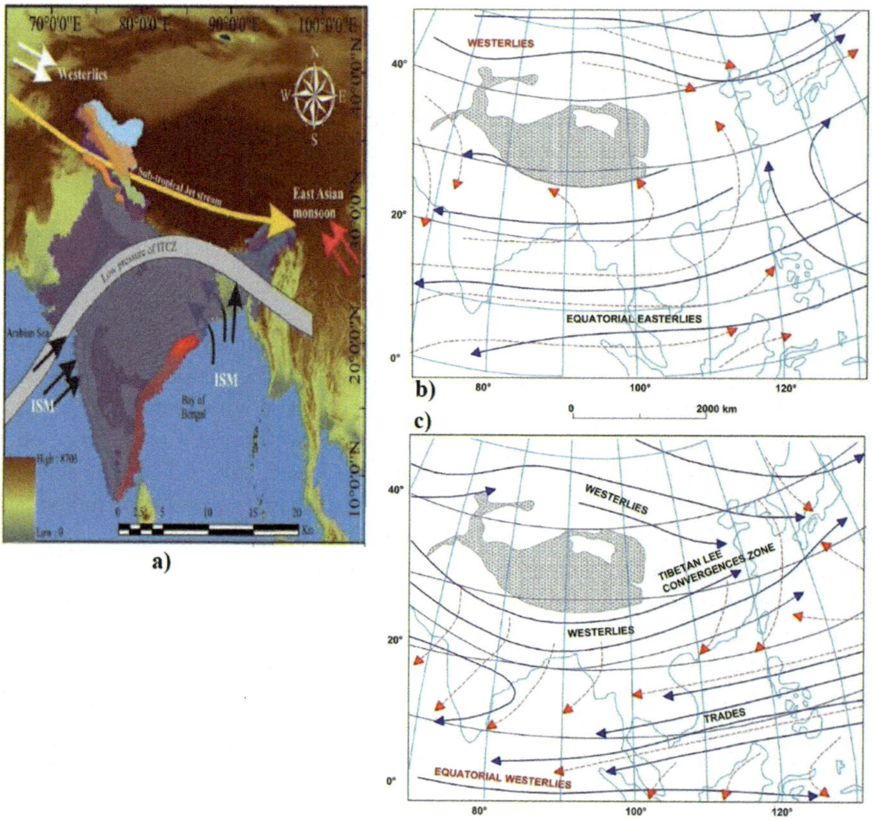


**Fig. 3.1** **a** Map showing path of major atmospheric winds influencing Himalaya in India (after Ali et al. 2018), **b** The characteristic air circulation over the western hemisphere showing the detailed circulation over southern and eastern Asia with respect to the Tibetan Plateau and bordering mountains in summer conditions and **c** winter conditions. The solid lines indicate airflow at about 3000 m and the dashed lines that at about 600 m. The names refer to the air systems. The area of Tibet and the Himalayas that lies above 6000 m is shaded (after Benn and Owen 1998)

## 3.3 Weather Mechanism: Dualistic Approach

The northwest Himalaya is one of the most magnificent mountain systems in the world and that is conspicuous by its giant ice fences housing large number of glaciers. The location and topographic setting, i.e. enclosed by huge mountain ranges of High Himalaya, Karakoram ranges, and Zanskar-Ladakh mountains in extreme northwest, and Pir Panjal and Dhauladhar ranges in southeast, support glaciation in higher altitudes.

The southeast segment of northwest Himalaya is a less effective barrier to summer monsoon as well as having less influence on the influx of the air mass moving from Westerly Trough to Himalaya. The influx of Westerly air mass is frequent in the High

Himalaya of Ladakh region in the vicinity of Pamir Pass during the winter season. In summer, the peaks of High Himalaya Mountain act as 'Warm Island' and centre for condensation of 'moist island'. Thus, the peaks serve as a conduit for bimodal distribution of air mass. The warm islands, through sublimation and coagulation, continuously release large quantities of latent heat into the upper stratum of the troposphere and the rising air columns maintain the powerful high pressure during summer season. There are two zones of high precipitation originating from two different sources of air masses, one from the local weather conditions and other from the Western Disturbances and monsoons. This precipitation usually takes place during the summer season particularly along the accumulation zone and helps in consolidating glaciers and reduces the effect of albedo (Azam et al. 2021).

An atmospheric disturbance known as Western Disturbance (WD) determines the precipitation over northwest Himalaya in winter. The western Himalaya Mountain blocks the Western Disturbances to cause maximum deterioration of weather conditions and results in abundant snowfall due to natural thermodynamic effect that leads to the development of subtropical glaciers in Himalaya. Although the periodicity and incidence of WD are variable, the impact on local weather is tremendous. The WD usually comes at an interval of 8–12 days having a short life of 5 days. Variations in summer monsoon and Western Disturbance have equally influenced the Himalayan glaciers (Hunt et al. 2018). Further, the Western Disturbance has been considered as a significant driver to influence the past and future changes in the climate of western Himalaya, particularly Ladakh, Zanskar and Karakoram. The studies suggest that the role of meteorological parameters in governing the snow cover extent and the annual changes in glacier mass balance are largely due to winter anomalies in accumulation, which in turn, are mainly due to the anomalies in precipitation and, to lesser extent, in temperature (Javed et al. 2023). The movement of the mid-latitudinal system predominantly in the westerly zonal flow causes plenty of weather disturbances across the Indian subcontinent. Western Himalaya during winter and pre-monsoon season receives precipitation, usually in the higher reaches, in the form of snow that plays a crucial role in the accumulation of snow/maintenance of glaciers. The westerly zonal flow is stronger and supports the mid-latitudinal system to reach the Indian subcontinent during ENSO (El Niño Southern Oscillation) years (Tamaddun et al. 2019). In other words, during ENSO years, it is likely that glaciers over the Himalaya will get more snow in winter and pre-monsoon seasons (Thomson and Vallis 2018).

The first record of comprehensive precipitation based on tree-ring studies indicates an increase in winter and spring precipitation during the recent century in northwest India, particularly in Uttarakhand (central Himalaya), owing to the Winter Monsoon (Singh et al. 2021). Based on modelling as well as paleoclimate studies, it has been demonstrated that Winter Monsoon plays a critical role in determining the climate of Himalaya. The term Western Disturbance was, therefore, replaced by Indian Winter Monsoon (Chevuturi et al. 2018). The glaciers of the Karakoram Mountain, particularly of Ladakh-Nubra region, stand out because of their divergent response and display of surge in glaciers, as opposed to the glaciers of other regions. Further, investigations were carried out, in terms of their intensity, precipitation/snowfall volume and wind speed, based on the study of satellite climate dataset of 39 winter seasons of Ladakh and its adjoining region. The data reveals that Karakoram Mountain witnessed a rise of 10.17% in the intensity of precipitation associated with Western Disturbance in recent decades and high percentage of snowfall amounting to 64.5%. Simultaneously, the contribution of snowfall from the sources other than Western Disturbance declined to 17.39% in recent decades (Dimri 2021). The enhanced intensity of Westerlies is found to be associated with change in increased upper troposphere baroclinic flow instability, shift in the mean latitudinal position of Subtropical Westerly Jet and enhance in carrying capacity of the moisture (Javed et al. 2022).

The tree-ring width consistency study at Padder in the semi-arid region of Kishtwar (Union Territory of Jammu and Kashmir), dating back to 1439AD, revealed low influence of summer monsoon (June–September) with − 16% contribution to annual precipitation, whereas a major portion of precipitation is brought by mid-latitudinal Westerlies in winter and spring (December–May). Chronology shows that highest correlation with SPI (Serial Peripheral Interface)-related exchange of data between two devices of April and May is attributed to the recent wetting and glacier expansion in northwest Himalaya and Karakoram (Yadav et al. 2017; Bahuguna et al. 2021). The recent wetting behaviour of northwest Himalayan glacier valleys and response of glaciers to climate change due to global warming is different from the other parts of the world. The extreme weather events, in other parts of the world, have resulted into the melting of ice sheets of Arctic, Antarctic and Greenland and calving of ice shelves producing icebergs and causing rise in the Greenland landmass (Fischer et al. 2021).

## 3.4 Slope Environment Lapse Rate: Characteristic of Temperature and Precipitation Variation

The regional mountain and local climate of glaciers is, unequivocally, influenced by the orographic processes and consequently the changes occurring to the glacial system impact the hydrology and ecosystem processes of the lower nival monsoon regime too. The glacier discharge, across the Himalayan arc and Hindu Kush region, after going through the nival, periglacial and monsoon processes, is in use in the downstream/lowland areas. The meteorological observatories installed in lower altitude, and rain-gauge/automatic weather stations installed recently in the high altitudes helps in generating the instrumental data, particularly with respect to rainfall/snowfall and temperature along with other meteorological parameters, to build more accurate models for the glacier regimes. The improved understanding of interaction between monsoon and nival-glacial systems resulted in improving the palaeoclimatic interpretations. The detailed analysis of primary weather elements, namely temperature and precipitation, for a prolonged period reveals that high mountain terrain exerts an important influence on local weather processes and induces temperature change with the gain of every hundred metres of altitude. There is large seasonal variation in mean maximum and mean minimum temperatures and precipitation. The variation in monthly mean maximum temperature at different stations is related to the variation in altitude. The large variation in minimum temperature within high mountain areas is due to the local topography of the sites, particularly in High Himalaya and trans-Himalaya. The monthly mean minimum temperature of altitudinal gradient ranges (Slope Environment Lapse rate—SEL) between 1 °C and 3.5 °C per 100 m. It is minimum during winter seasons of November to May in Ladakh region and winter season of December to February in Pir Panjal, Dhauladhar and central Himalaya. The winter months experience larger precipitation. The SEL between − 8.5 and 10.2 °C per 100 m is highest during winter in Ladakh-Zanskar region compared to Lesser Himalayan ranges like Pir Panjal and Dhauladhar due to highest relative relief of High Himalaya. The night temperature fluctuates between 10 and 20 °C from the normal, generally during November to May in High Himalayan region. The winter precipitation in the form of snowfall ranges between 5.5 and 10 m from lower altitude to High Himalayan glacier snouts (Suru-Zanskar). The winter precipitation in Lesser Himalayan (Pir Panjal and Dhauladhar ranges) glacier valleys ranges between 1 and 4 m. The eastern Himalaya receives the highest annual precipitation of 2800 mm from June to September in the form of rainfall (89%), and Karakoram range receives a total annual precipitation of 584 mm mostly in the form of snowfall (81%) (Azam et al. 2021). The western Himalayan glaciers contribute 60.7% of annual flow to the major rivers of northern India in the form of melt water, whereas eastern Himalayan glaciers contribute, relatively less due to high snowline and intense monsoon (Kulkarni et al. 2021). Studies show that Indus basin has a higher dependence on snow and glacier melt (44.7%), followed by Brahmaputra (15%) and Ganga (13.4%) basins (Khanal et al. 202).

## 3.5 Northwest Himalaya

The glaciers in the northwest Himalaya are mostly sustained by Indian Summer Monsoon system, which is rooted in phenomenon of the Inter-Tropical Convergence Zone (ITCZ) that arises because of the seasonal differences in temperature and pressure of Northern Hemisphere and Southern Hemisphere. In particular, the temperature gradient between the Himalaya-Tibet (orogen) massif and surrounding oceans drives summer monsoon system. The elevated region acts as orographic barrier that influences atmospheric circulation pattern and plays a vital role in modifying strength of South Asian monsoon (Wang et al. 2022). The high elevated region of Himalaya-Tibetan orogen acts as mid-troposphere source of heat during the summer and as a heat sink during the winter. The winter-spring snow cover over Himalaya-Tibet massif plays key role in the genesis of the thermal contrast. The Western and trans-Himalaya region receives precipitation mostly during winter months (November–April) due to Western Disturbances (WD) that originates in the Mediterranean Sea. These varying weather systems give rise to precipitation gradients along and across the Himalaya due to mechanical and thermal forcing. The relative importance of varying weather systems in bringing moisture to the peaks of the Himalaya and trans-Himalaya lies in defining the spatial and temporal distribution of glaciers, in past and present. Both circulation systems follow pronounced seasonal cycle that leads to distinct annual peak in temperature and precipitation distribution (Krishnan et al. 2019a).

Global studies show that, during the last century, glaciers in mid-latitude have tendency to retreat with small-sized glaciers disappearing and large-sized shrinking and fragmenting slowly, whereas the glaciers in the polar regions are warming considerably. Scientific study during last three decades on Kolahoi-Shishram in Liddar valley (Kashmir, Union Territory of Jammu and Kashmir) and Naradu in Baspa basin (upper Sutlej valley, Himachal Pradesh) glaciers in northwest Himalaya, and Machoi, Durung Drung, Kangriz (Upper Indus basin, Ladakh) and Siachen (Nubra-Shyok, Ladakh) glaciers in west Himalaya (Fig. 3.2) are presented here as case studies, in the background of global studies, to underline the impact of climate change and heat dome effect on glacier dynamics in terms of variation in length, mass balance and impact of micro-climate variables.

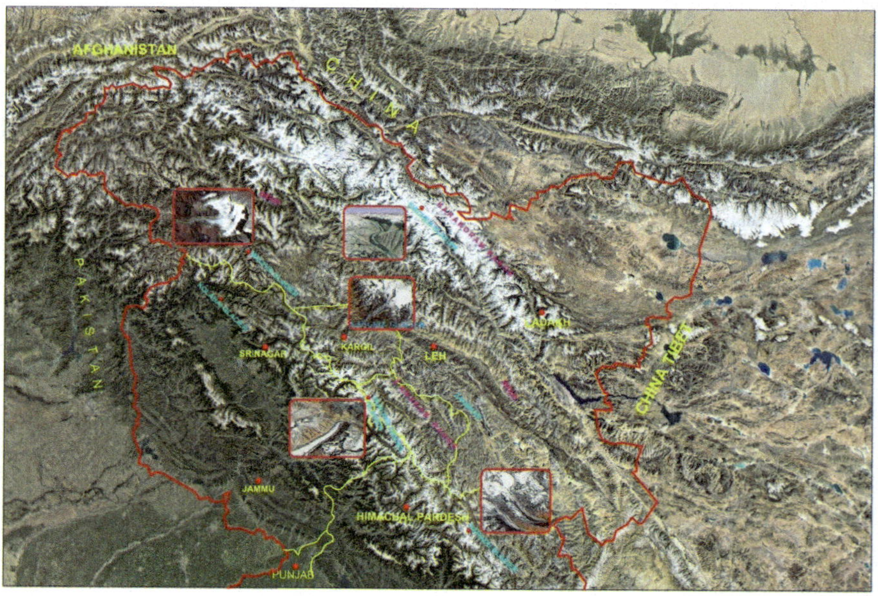


**Fig. 3.2** Physiographic map of northwest Himalaya demarcating boundaries of Union Territories of Jammu and Kashmir, and Ladakh and state of Himachal Pradesh (India) and locations of Siachen, Machoi, Durung Drung, Kangriz/Parkachik, Kolahoi and Naradu glaciers are marked

### 3.5.1 *Liddar Valley (Pir Panjal Range)*

Liddar valley occupies the southern part of Kashmir Himalayan synclinorium and forms part of Pir Panjal range. It is girdled on three sides by lofty ridges, the Saribal-Katsal ridge (4800 m amsl) in the east, Wokhbal (4200 m amsl) in the west and Basmai-Kazimpathlinbal (4800 m amsl) in the north. The interior of northern part of the Liddar valley has concentration of ridges which runs parallel to the course of the river. The distinct climatic characteristic of the valley is due to its location at higher altitude and physiographic setting which means that it is enclosed on all sides by mountain ranges. The region has a sub-Mediterranean type of climate with nearly 80% of the annual rainfall concentrated during winter and spring months. The region has humid continental, severe winter, moist all seasons, short warm summer (Dfb) climate regime (Indian Meteorological Department 2014). The cold arid type of climate and the rugged topography of high-altitude regions of Himalaya serves as a handicap in the data collection due to inaccessible and inhospitable terrain. Even the precipitation and temperature data of Lesser Himalaya, such as Pir Panjal (Kashmir) and Kinnaur, Dhauladhar (Himachal Pradesh), is available from rain-gauge stations only. Most of the rain-gauge stations are located at the elevations well below 2000 m amsl that are accessible, whereas very few stations are located at high altitudes. The paucity of meteorological data from high-altitude rugged terrain is redeemed by the satellite data information. However, on the ground, verification of the information obtained through satellites of high rugged terrain is difficult. The meteorological observatories at headquarters of the Union Territory of Jammu and Kashmir (Srinagar and Jammu), Union Territory of Ladakh (Leh) and Himachal Pradesh (Shimla) are the only source of instrumental data throughout the year. Detailed analysis of the data on primary weather elements (temperature and precipitation) for a limited period, therefore, is used to derive gradient index of temperature and precipitation (snowfall and rainfall) at different altitudes of mountain terrain to assess changes, after every 100 m of elevation, in temperature and rainfall/snowfall.

The large seasonal variation in the mean maximum and mean minimum temperatures and precipitation, and consistency in variation in monthly mean maximum temperatures at different stations is related to the variation in altitude. The variation in minimum temperatures is large in high mountain areas and subject to the local topography of the site. The temperature variation suggests marked seasonality in weather, with long cold season lasting for about six months but a short summer season. The winter receives heavy precipitation caused by Western Disturbances. The normal daily maximum temperature is between 2 and 20 °C during November to May. The mean minimum temperature ranges between − 13.2 °C and + 1 °C. The temperature distribution in high altitude shows characteristic altitudinal gradient. The variability of monthly mean maximum of the altitudinal gradient is large with a standard deviation of 0.9° to 1.8° per 100 m (Table 3.1) between Pahalgam (2200 m amsl) and Pharsulan (2300 m amsl), Aru (2400 m amsl) and Chandanwari (2700 m amsl), respectively.

**Table 3.1** Mean monthly altitudinal gradient of temperature (°C/100 m) of selected sites of the northern part of the Liddar valley with reference to Pahalgam (2000 m amsl) from 1973 to 1988 (after Kaul 1990a)

| Month | Pharsulan (2300 m amsl) | | Aru (2400 m amsl) | | Chandanwari (2700 m amsl) | |
|---|---|---|---|---|---|---|
| | Max | Min | Max | Min | Max | Min |
| Jan | 0.9 | 2.3 | 1.0 | 1.2 | 0.6 | 0.7 |
| Feb | 1.2 | 2.1 | 1.1 | 1.8 | 0.7 | 0.8 |
| Mar | 1.8 | 1.2 | 1.2 | 0.8 | 0.5 | 1.0 |
| Apr | 1.2 | 0.9 | 1.3 | 0.6 | 0.6 | 0.5 |
| May | 0.9 | 1.1 | 1.o | 0.7 | 0.6 | 0.4 |
| Jun | 1.0 | 0.9 | 1.8 | 0.2 | 1.1 | 0.3 |
| Jul | 3.2 | 2.5 | 2.2 | 1.9 | 1.2 | 0.4 |
| Aug | 3.6 | 3.5 | 2.3 | 2.2 | 1.3 | 1.0 |
| Sep | 3.9 | 2.5 | 2.4 | 1.5 | 1.2 | 1.0 |
| Oct | 3.9 | 0.5 | 2.7 | 2.2 | 1.4 | 1.1 |
| Nov | 1.7 | 0.9 | 1.2 | 1.3 | 0.8 | 0.7 |
| Dec | 1.5 | 1.4 | 1.0 | 1.9 | 0.4 | 0.9 |

*Source* Indian Meteorological Department, Srinagar, Jammu & Kashmir

The winter precipitation during November to March is received in the form of snowfall and the pattern varies with the altitude. As recorded in the field study during 1983–1985, the winter precipitation ranges between 1 and 4 m from Pahalgam to Kolahoi-Shishram (3800 m amsl), respectively. The altitude gradient of monthly mean minimum temperature during the months of December to February ranges between 1 and 3.5 °C per 100 m. The winter months at Chandanwari, Aru, and Pharsulan/Pahalgam experience heavy solid precipitation and receive 820, 730 and 580 mm equivalent to rainfall.

Pahalgam and Gulmarg stations in the Kashmir valley, located at an elevation of ~ 2500 m amsl, registered significant increase in average annual temperature by 1.13 and 1.04 °C during 1980–2016, respectively, whereas Kokernag and Kupwara at an elevation of ~ 1800–2000 m amsl show increase by 0.9 and 1.0 °C, respectively, and an increase of 0.65 °C and 0.44 °C is recorded from Srinagar and Qazigund at an elevation of ~ 1700–1600 m amsl, respectively (Zaz et al. 2019). Further, Pahalgam and Kupwara showed highest rise of ~ 1.3 °C followed by Kokernag of 1.2 °C and Srinagar of 1.1 °C, whereas Gulmarg and Qazigund showed less than 0.6 °C. On the other hand, there is low increase of 0.3 °C in minimum temperature at Srinagar and highest of 1.2 °C at Gulmarg.

The temperature during summer (June–October) ranges between 16.5 and 30 °C. The mean monthly maximum temperature of altitude gradient varies between 1 and 3.5 °C per 100 m with maximum of 4 °C from August to October. The altitudinal gradient of mean monthly minimum temperature ranges between 1 and 2 °C per 100 m. The rainfall variation with altitude is same during summer and winter.

In lower altitude, the rainfall variation is highest (194 mm) at Chandanwari and lowest (142 mm) at Pahalgam (Table 3.2). The rainfall gets marginal benefit from the monsoon because of the rain-shadow effect of Pir Panjal and due to local conditions.

The deforestation has escalated concentration of greenhouse gasses in the atmosphere in last two decades leading to the increase in mean monthly maximum temperature during winter and summer with decrease in the winter precipitation accompanied by events of extreme rainfall. The average annual temperature shows increase at all stations of Kashmir Himalaya although with different magnitudes and patterns under different Representative Concentration Pathways (RCP). The simulation reveals that climate in the region may progressively get warmer in future by an increment of 0.36–1.48 °C, and seasonal and annual precipitation is likely to decrease as it shows spatially incoherent trend for cold temperature-based indices. Precipitation indices, on the other hand, show decrease in most of the area with some increase in higher altitudes due to local orographic control (Ahsan et al. 2022).

### *3.5.2 Naradu Glacier Valley (Dhauladhar Range)*

Naradu glacier valley in Dhauladhar range of Himalaya, district Kinnaur, Himachal Pradesh, is located on the bank of Baspa River near village Chitkul on Hindustan-Tibet Highway. The glacier is in the upper reaches of Naradu Garang valley enclosed in south and southeast by high mountain peaks, namely Khimsung (5600 m amsl), Khomsung (5700 m amsl) and Khimloga (5400 m amsl). The south boundary of the valley serves as a major water divide between the first-order basins of Sutlej and Ganga in India. Naradu glacier is one among 89 glaciers of Baspa basin that forms the $V^{th}$ order basin of Sutlej River.

Naradu glacier is a dirty glacier covered with debris and the valley has distinct climate characteristics due to its location in higher altitude and topographic setting of being enclosed by higher Himalaya and steep Lesser Himalayan range. The valley experiences diverse climate conditions ranging from cold arctic in the upper, alpine in the middle and humid in the lower segments. Nearly 70% of the annual precipitation is concentrated to the winter season. The precipitation and temperature data are available from only two rain-gauge stations installed at Chitkul and Sangla. The observatory at Shimla (at an aerial distance of about 100 kms) is the only one to record the meteorological data throughout the year. The analysis of main weather parameters, i.e. temperature and precipitation, recorded at Chitkul and Sangla at Naradu valley, from the year 1994 to 2003, shows that high mountainous terrain exerts an influence on local weather processes. The temperature and precipitation changes after every 100 m of the altitude show long seasonal and diurnal variations in mean maximum and minimum temperatures at different altitudes within the basin. The variations are attributed to local topography, aspects or orientation of slopes with a particular reference to southwest and northeast direction, reflecting that the basin is dominated in the winter season by Westerlies. The basin experiences marked seasonality in weather with a long cold season lasting for seven months and a short

**Table 3.2** Mean monthly temperature and rainfall of the selected sites of northern part of the Liddar valley from 1973 to 1988 (after Kaul 1990a)

| Month | Pahalgam (2200 m amsl) | | | Pharsulan (2300 m amsl) | | | Aru (2400 m amsl) | | | Chandanwari (2700 m amsl) | | |
|---|---|---|---|---|---|---|---|---|---|---|---|---|
| | Temp. (°C) | | Rainfall (mm) | Temp. (°C) | | Rainfal l (mm) | Temp. (°C) | | Rainfal l (mm) | Temp. (°C) | | Rainfall (mm) |
| | Max | Min | | Ma x | Min | | Max | Min | | Max | Min | |
| Jan | 5.1 | – 8.6 | 122 | 4.2 | – 10.9 | 135 | 1.9 | – 12.3 | 172 | 3.1 | – 11.7 | 146 |
| Feb | 6.8 | – 9.1 | 72 | 5.6 | – 11.3 | 82 | 3.1 | – 13.2 | 84 | 4.2 | – 12.6 | 92 |
| Mar | 14.7 | – 2.3 | 124 | 12.9 | – 3.5 | 136 | 11.9 | – 4.7 | 153 | 12.2 | – 4.0 | 126 |
| Apr | 20.0 | 2.9 | 96 | 18.8 | 2.0 | 105 | 16.8 | 1.1 | 122 | 17.7 | 1.7 | 100 |
| May | 24.8 | 4.9 | 78 | 23.9 | 3.8 | 94 | 20.9 | 3.2 | 105 | 25.1 | 3.6 | 98 |
| Jun | 30.0 | 5.1 | 142 | 29.0 | 4.2 | 167 | 24.6 | 3.9 | 194 | 26.4 | 4.7 | 170 |
| Jul | 29.5 | 10.7 | 68 | 26.3 | 8.2 | 82 | 23.4 | 5.8 | 99 | 25.2 | 6.8 | 88 |
| Aug | 28.3 | 10.5 | 64 | 24.7 | 7.6 | 73 | 22.2 | 5.6 | 86 | 23.8 | 6.1 | 74 |
| Sep | 26.2 | 8.2 | 16 | 22.3 | 5.7 | 18 | 202 | 4.9 | 34 | 21.4 | 5.2 | 17 |
| Oct | 22.3 | 0.9 | 11 | 17.1 | – 1.4 | 14 | 15.4 | – 4.0 | 16 | 16.9 | – 3.2 | 15 |
| Nov | 14.1 | – 4.9 | 4 | 12.4 | – 5.8 | 6 | 9.8 | 8.6 | 8 | 11.7 | – 7.1 | 6 |
| Dec | 6.4 | – 6.7 | 88 | 4.9 | – 8.1 | 115 | 2.2 | – 11.2 | 164 | 4.5 | – 10.4 | 122 |

*Source* Indian Meteorological Department, Srinagar, Jammu and Kashmir

summer season for five months. The Naradu valley, based on seasonal variations, experiences two seasons.

Two meteorological observatories were set up, one near the snout of the Naradu glacier and other at firn line of the Naradu glacier, to record the annual variation in weather parameters and assess the impact on the glacier dynamics and lapse rate of temperature and precipitation that forms the characteristics of cold arid mountain slope of Naradu Garang valley. The automatic weather system (AWS, Model DT50-Earth and Atmospheric Sciences, USA), procured under the funds received from Department of Science and Technology (DST), Government of India, was mounted on galvanized steel tower with A-50 data logger and package of sensors operating on solar batteries (a set of 12 batteries) and housed in weatherproof enclosure. It was installed at an altitude of 4100 m amsl in the year 1994 at the base camp of Naradu glacier valley. The AWS comprised sensors to record air temperature, humidity, speed and direction of wind, cloud, precipitation (snowfall and rainfall), solar radiation, sunshine hours and albedo (Koul and Ganjoo 2010). The weather station had a resolution of 0.1 °C with an accuracy of ± 1% of full scale and recorded the data on everyday basis between 1994 and 2003. A manual observatory was also set up in the year 2000 at an altitude of 4950 m amsl in the vicinity of firn line of the Naradu glacier to record changes in temperature and precipitation during the ablation season from May to October.

## 3.6 Meteorological Analyses of Naradu Glacier Valley

The micro-meteorological parameters, such as temperature, precipitation, solar radiation and sunshine hours, were recorded with the help of Automatic Weather Station during the years from 1994 to 1997, and from 2000 to 2003 (Fig. 3.3a–c) for critical analyses. The analyses of micro-meteorological parameters reveal noticeable diurnal and seasonal variations to define the climate of Naradu glacier valley.

### *3.6.1 Temperature*

The difference between the mean daily maximum and minimum diurnal temperature is 8–14 °C during the summer season and – 10 to – 3 °C during the winter season. The air temperature increases from the month of May to September with July being the hottest when mercury touches the mark of 21.9 °C. The air temperature, thereafter, begins to decrease from the month of mid-September to April, with February being the coldest when temperature dips to as low as – 18.9 °C. The air temperature reaches to maximum during the day-time and falls to the lowest at the time of sunrise. The analyses of the temperature regime recorded from the year 1994–2003 indicate that the summers in Naradu valley are, on the whole, warmer and the average maximum temperature during the summer season ranges from 14 to 16 °C. The monthly diurnal

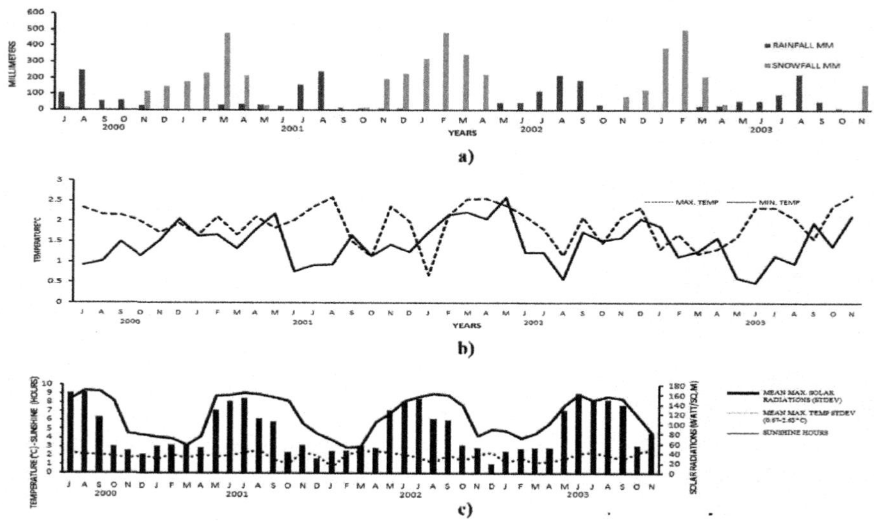


**Fig. 3.3** Representation of various meteorological parameters recorded at Naradu glacier base camp from 2000 to 2003 **a** monthly precipitation, **b** mean maximum and minimum temperature (stdev), and **c** comparative graph of temperature, solar radiation and sunshine hours,

temperature range during summer and winter shows an increasing trend. The average diurnal range of summer is 15.2 °C whereas it is 14 °C in winter. The highest diurnal range is seen in the months of March (18.4 °C) and September (16.4 °C). There is marked seasonality in mean monthly maximum and minimum temperature. The mean minimum temperature during winter (October to April) falls to nearly –21.0 °C. The average maximum temperature during the season ranges from –1.62 to 9.67 °C and average minimum temperature between –3.61 and –18.49 °C. The study of temperature data recorded at Chitkul and Sangla in Naradu valley, from the year 1994–2003, reveals that there is a steep fall, dipping as low as –18.9 °C, in the night temperature during the winter season. The night temperatures deviate by 10–20 °C, in general, from the normal during the months from November to March. The months from December to February experience peak winter with the month of February as the coldest when mean average minimum temperature falls to – 18.9 °C. Frost is prevalent during the winter season. Abrupt fall in temperature in higher altitude leads to permafrost conditions and facilitates periglacial landscape above the altitude of 4400 m amsl in the Naradu valley. The mean maximum temperature varies from 9.7 to – 6.3 °C in accumulation zone of glacier near ELA and from 16.3 to 5.4 °C near snout during summer season (Table 3.3).

The summer season in the Naradu valley is short and lasts from the months of May to September. The temperature curve for the mean monthly maximum temperature ranges between 12.18 and 21.0 °C. The monthly minimum average temperature fluctuates between –1.63 and 7.12 °C. The temperature data recorded at Chitkul and Sangla, from the years 1994 to 2003, reveals that the months of June and July are the hottest with an average maximum temperature reaching to 21.7 °C. The

**Table 3.3** Comparative statement of temperature range at snout and equilibrium line of Naradu glacier

| Temperature range (in °C) near equilibrium line (4950 m amsl) | | | | |
|---|---|---|---|---|
| Year | 2000 | 2001 | 2002 | 2003 |
| Max. temp | 8.9 (Sept.5) | 9.8 (July 11) | 10.2(July 17) | 7.8(Aug. 31) |
| Min. temp | – 8.2(Sept. 17) | – 5.0 (July 25) | – 5.8(July 28) | – 6.3(Aug. 21) |
| Range | 17.1 | 14.6 | 16 | 14.1 |
| Mean 15.5 | | | | |
| Mean Diurnal Range | 13.8 | 11.7 | 13.2 | 9.4 |
| Mean 12.02 | | | | |
| *Temperature range (in °C) near glacier snout (4395 m amsl)* | | | | |
| Year | 2000 | 2001 | 2002 | 2003 |
| Max. temp | 18.9 (July 4) | 21.8 (July 17) | 19.9 (July 11) | 19.6 (Aug 5) |
| Min. temp | 7.6 (July 22) | 9.5 (July 28) | 5.8 (July 13) | 5.4 (Aug 19) |
| Range | 11.3 | 13.3 | 14.1 | 14.2 |
| Mean 13.2 | | | | |
| Mean Diurnal Range | 14.8 | 13.8 | 14.2 | 14.5 |
| Mean 14.3 | | | | |

diurnal variation in temperature is large during summer season facilitating physical weathering due to freeze and thaw action. As a result, the valley walls experience frost shattering and serve a major source for the supraglacial debris.

### 3.6.2 Precipitation

The analyses of the precipitation data collected during the years from 1994 to 1997 and again from 2000 to 2003 reveal that highest precipitation occurred in the year 2002 amounting to 2.23 m (88.10 inches) followed by 2.20 m (86.90 inches) in the year 1997 and 2.17 m (85.75 inches) in the year 1996. Precipitation received during the years 2001, 1995 and 2003 was 2.12 m (83.85 inches), 1.91 m (75.3 inches) and 1.84 m (72.61 inches), respectively. About two-third of the total precipitation was received in the form of snowfall. The snowfall was received mostly between the months of October and April during the winter season. The snowfall was recorded to 1.80 m (71.1 inches) in 2001–2002, 1.36 m (53.63 inches) in 1996–1997 and 0.99 m (39.04 inches) in 1995–1996. The snowfall pattern varied considerably during the winter season with maximum of 0.87 m (34.45 inches) snowfall occurring in the months of January, February and March.

The rainfall pattern in the Naradu glacier valley also displays high degree of variation. The highest rainfall was received in the year 1997 as compared to other years. The highest rainfall of 17.6inches was in the month of August. The total rainfall received during the years 1997,1996, 1995, 2002, 2001 and 2003 was 1.24 m (48.9 inches), 1 m (39.72 inches), 0.95 m (37.78 inches), 0.65 m (25.61 inches), 0.56 m (22.3 inches) and 0.55 m (21.66 inches), respectively. The precipitation is primarily received in the form of rain during the summer season, whereas occasional snowfall occurs during the summer season in the upper reaches of the glacier valley. The meteorological data reveal that maximum precipitation of 0.18 m (7.33 inches) and 0.28 m (11.29 inches) during the summer season occurs in the months of July and August, respectively. The rainfall in the month of August has considerable geomorphological significance as it causes high discharge in the stream of Naradu Garang and the tributaries of Baspa River, and consequently brings down high quantity of debris from the mountain slopes. The high impact of monsoon on the slopes of low altitude areas results into increase in the intensity of rainfall, caused by cloud bursts. The high-intensity rainfall has been responsible to cause catastrophic sliding in Sangla and Wangtu segments of Sutlej River basin leading to the blockade of rivers at places. During last two decades, severe summer monsoon has caused blockade of Sutlej River as was observed in 1996 when the Sangla valley and upper reaches of the region were cut off from rest of the country for nearly five months due to the blockade of Sutlej River at Wangtu.

### *3.6.3 Solar Radiation*

The amount of solar radiation received at a place varies with respect to atmospheric conditions and the season. A part of solar radiation falling on any surface is reflected and as such certain amount of energy is lost, known as albedo. The analysis of solar radiation data reveals that solar radiation reached up to 625 watts/m$^2$. The maximum day-time insulation from the months of May to October ranges between 220 and 625 watts/m$^2$. The highest maximum solar radiation of 962 watts/m$^2$ is received during the months of July and August, whereas the lowest minimum of 163 watts/m$^2$ is received in the months of January and February. The mean monthly maximum solar radiation for the months of June to September ranged between 590 and 490 watts/m$^2$ for the years 2000 and 2001 as compared to 538watts/m$^2$ and 454watts/m$^2$ for the years 2002 and 2003. During the summer season, from the months of May to September, the solar radiation ranged between 402 and 620 watts/m$^2$. The average solar radiation during the winter months (November–April) ranged from 220 to 400 watts/m$^2$.

### *3.6.4 Sunshine Hours*

The duration of sunshine is a powerful indicator to assess the degree-day melting of glaciers. The sunshine hours ranged between 6.49 and 10.75 h/day from the months of May to October. Peak sunshine hours were recorded during the months of June, July and August. During the cold months of November to April, the sunshine hours ranged from 2.89 to 6.51 h/day. The mean sunshine hours during the winters ranged between 3.6 and 4.5 h/day. The average sunshine of 8.3 h/day in the years of 2000 and 2001 is higher than the sunshine hours of 7.6 h/day received in the years of 2002 and 2003.

### *3.6.5 Wind*

The wind flows down the slope in southwest and southeast direction from Khimloga and Khomsung Pass to Baspa River valley at a speed of 3–10 km/h during the summer season, and between 5 and 15 km/h during the winter season. The direction of wind occasionally changes to north and northwest. The wind is calm in the morning and gains momentum by noon and afternoon. The wind, once again, is calm in the evening during the summer season, but gains momentum during the night in the winter season.

## 3.7 West Himalaya

Ladakh is a mountainous part of the Indian subcontinent with high mountain ranges, elevated plateau and rocky gorges that is surrounded by lofty ranges of High Himalaya comprising towering peaks of Nun (7135 m amsl) and Kun (7087 m amsl). The Karakoram ranges house large number of glaciers that are confined in protracted zones of the mountains. The glacier valleys have steep slopes and relative relief of 2500 m in 3–4 km in the main valley, whereas the relative relief in tributary valleys is 200 m in 5 km. The magnitude of relief and overall steepness of slopes provide suitable conditions for semi-arid to arid type of environment in the region. The aridity in West Himalaya increases with altitude and the areas of West Himalaya that are in the rain shadow also experience aridity due to high mountain masses. The climate of the area in West Himalaya is similar to that of Central Asia. Lack of meteorological data for different glacier basins, such as Durung Drung (Zanskar), Machoi (Drass) and Siachen (Nubra), serves handicap to assess annual variation in the weather parameters and understand the impact of climate on glacier dynamics. However, long-term temperature and precipitation rain-gauge records are available for the Kargil (2704 m amsl), Leh (3513 m amsl) and Drass (3066 m amsl) stations of Ladakh region (Table 3.4). All the stations are far away from the snouts of glaciers and are located at the lower peripheries of glacier basins.

**Table 3.4** Mean monthly (min. and max.) temperature (in °C) of Leh, Kargil and Drass stations

| Months | Leh | | Kargil | | Drass | |
|---|---|---|---|---|---|---|
| | Max | Min | Max | Min | Max | Min |
| January | − 2.80 | − 14.00 | − 4.20 | − 13.35 | − 6.14 | − 21.11 |
| February | 0.80 | − 11.85 | − 1.65 | − 12.15 | − 2.44 | − 19.46 |
| March | 6.40 | − 6.35 | 4.65 | − 5.35 | 2.21 | − 14.36 |
| April | 12.35 | − 1.20 | 14.00 | 3.40 | 6.98 | − 7.61 |
| May | 17.10 | 2.75 | 21.60 | 9.35 | 11.83 | − 3.34 |
| June | 21.10 | 6.65 | 24.70 | 13.35 | 19.57 | − 3.48 |
| July | 24.70 | 10.15 | 29.70 | 17.70 | 19.45 | 0.62 |
| August | 24.20 | 9.60 | 28.80 | 17.20 | 16.40 | 0.65 |
| September | 20.85 | 5.40 | 24.85 | 12.50 | 12.68 | − 2.89 |
| October | 14.15 | − 0.90 | 18.50 | 5.40 | 12.96 | − 5.20 |
| November | 7.8 | − 6.60 | 10.35 | − 1.30 | 5.73 | − 8.91 |
| December | 1.6 | − 11.15 | 1.2 | − 7.95 | − 1.76 | − 15.50 |

### *3.7.1 Climate Characteristics of Ladakh (Leh-Nubra-Kargil-Drass-Suru-Zanskar) Region*

The western Himalaya is a complex mountain system of Suru-Zanskar-Drass-Karakoram (Union Territory of Ladakh) that acts as an effective barrier to Westerlies air flow and Western Disturbances (originating from Mediterranean and Caspian Ocean) moving aloft Pamir ranges. They are moisture laden extratropical cyclone in the upper atmosphere, unlike their tropical counterparts where moisture is carried in the lower atmosphere. These air masses have a southward trajectory and get enclosed in southward propagating subtropical Westerly Jet to produce extreme precipitation that enhances over Himalaya due to orographic-land–atmosphere interactions (Liu et al. 2020). The local climate of the glaciers, which forces mass and energy exchange on the glaciers, is primarily driven by Himalaya-Tibetan Plateau orography and its alignment influences the high-altitude climate as noticed in the distribution of precipitation in northwest Himalaya (Zhuang et al. 2022). Most of the glaciers in Zanskar, Suru, Drass and Nubra valleys in Ladakh are north facing due to which they have large ice cover compared to the south facing glaciers that have less ice cover because they receive high amount of solar radiation. Further, glaciers of the region are confined to the shadow zones with steep slopes (25° and 45°) in the altitudinal range from 4600 to 5400 m amsl.

The present study deals with the monitoring of weather in the background of climate change and deciphering role of orography to the changes occurring to the surface conditions of the glaciers, e.g. snow cover duration. In view of the above, the AWS procured under the DST-funded project was installed at Durung Drung glacier,

Zanskar at an altitude of 4150 m amsl and data was recorded for the field seasons spread from June to October for the years from 2005 to 2008.

The data generated by Henry Osmaston at Stongde and Padum (Zanskar) on temperature and precipitation patterns at different altitude level, for the summer seasons for the years from 2005 to 2007, was also consulted. The data is corroborated with the meteorological data of Indian Meteorological Department (IMD). Suru-Zanskar valley, confined to the lap of trans-Himalaya terrain, has a distinct seasonal characteristic related to Western Disturbances and local micro-meteorological parameters. During past one decade, researchers have been using long-term meteorological satellite data sets like Modern- Era Retrospective Analysis, version 2 (MERRA-2) and CRU-TS (4.02) for the analysis of annual mean temperature and precipitation trends. The research team of Wadia Institute of Himalayan Geology, Dehradun (India), used satellite time series data (CRU-TS 4.02) for the period between 1901 and 2017 to highlight the distinct pattern of change in the climate parameters. The analysis of data reveals that annual mean temperature and precipitation have been 0.99 ± 0.45 °C and 393 ± 76 mm, respectively. The local climate variability reveals that Suru sub-basin, Zanskar, has become warmer (0.96 °C) and wetter (445 mm) (Shukla et al. 2020b). However, conflicting signals of change in climate, in terms of variation in snowfall and extent of snow cover, is reported from Zanskar valley on the basis of satellite data (version MOD10A2) of National Snow and Ice Data Centre, USA. The analysis of the data of last two decades further reveals that annual average snow cover in Zanskar valley is 68%, with maximum of 96% in the month of March and minimum of 32% in the month of August. The highest amount of snowfall and extent of snow cover was recorded in 2015 (80%), whereas the lowest was recorded in 2007 (59%). Trendline analysis shows an overall positive trend in the percentage of snowfall (precipitation) from the year 2004 onwards (Khali et al. 2022). The satellite data corroborate well with the IMD data for Ladakh and Kargil-Drass with respect to primary trends of weather elements.

The detailed analyses of primary weather data elements (temperature, precipitation, humidity, etc.) for Leh, Kargil and Drass for a period of forty years (1975–2015) show that high mountain terrain exerts an important influence on micro-climatic and weather processes. The temperature and precipitation regimes show variation and change after every 100 m of altitude (Table 3.5). There is consistency in variation in mean monthly maximum and minimum temperature at different stations of Leh, Kargil and Drass. The winter season, extending for a long duration from the months of October to May, in Ladakh and Suru-Zanskar valley is severely cold, extremely dry and chilly. The temperature falls to – 31.11 °C during the winter season. The mean monthly maximum temperature during this season ranges from 17.10 °C to −2.80 °C at Leh, 21.60 °C to – 4.20 °C at Kargil and 11.83 °C to – 6.14 °C at Drass (Fig. 3.4a–d). The temperature starts to fall below freezing point from the month of October onwards and the mean monthly minimum temperature varies from – 14 to – 0.90 °C at Leh, – 13.35 to 9.35 °C at Kargil and −31.11 to −3.34 °C at Drass. The peak winter is from the months of December to February with January the coldest, when mean monthly temperature dips to −31.11 °C at Drass. The temperature starts ameliorating from the end of April. The transition from winter to summer season commences from the month of May, as the temperature records visible improvement. Bright sunny days and cool nights are characteristic of the month and freeze and thaw process prevails at low altitudes up to 4000 m amsl. An analysis of maximum and minimum temperatures of all three stations show a wide range of fluctuation in the winter season due to the Western Disturbances. Abrupt fall in the night temperature is a peculiar characteristic of winter months.

In view of the above characteristics of weather conditions of Leh, Kargil and Drass, the seasons of a year can be divided into two, namely long winter season from the months of October to May, and short summer season from the months of June to September. The precipitation, in the form of snowfall, during the winter season commences from the month of November and a highest number of snowfall occurrences take place between the months of December and April. Leh receives 49.03% of precipitation in the winter season (Table 3.6). Kargil which is at lower altitude than Leh receives 88.95% of the precipitation from the month of October to May (Table 3.7). During the winter season, the highest percentage of precipitation (91.31%) is received over western Himalaya including Drass (Table 3.8) (Fig. 3.3.a–d) due to its location at the trajectory of Western Disturbance (Hunt et al. 2024). The El Niño Southern Oscillation (ENSO) also activates the Westerlies in the study region during summer period causing intense rainfall near the snouts and snowfall in the equilibrium line of the glaciers. The region receives summer precipitation primarily in the form of rain and occasionally in the form of snow in the upper reaches of mountains and glacier valleys. The rainfall data at Leh, Kargil and Drass reveal that

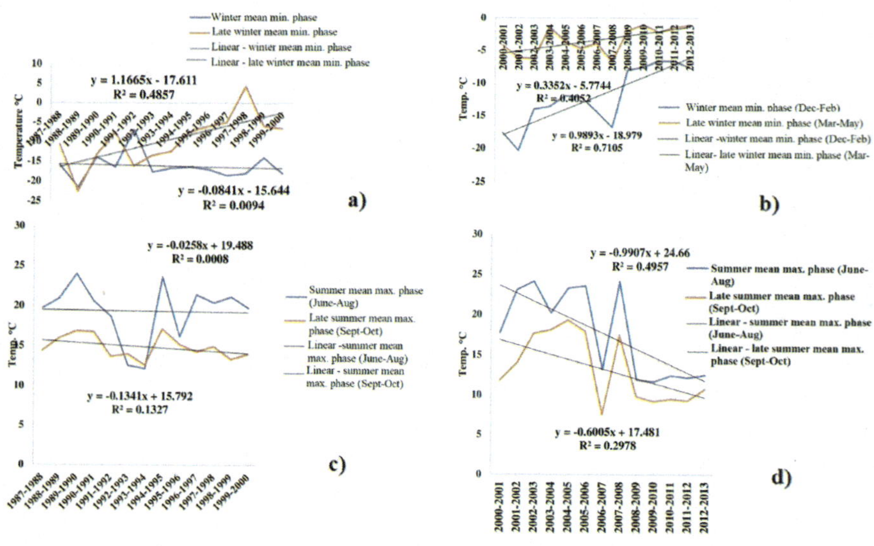


**Fig. 3.4** Graphic representation of trendlines of **a** mean minimum temperature of winter season from 1987 to 2000, **b** mean minimum temperature of winter season from 2000 to 2013, **c** mean maximum temperature of summer season from 1987 to 2000, and **d** mean maximum temperature of summer season from 2000 to 2013 at Drass (Ladakh)

**Table 3.5** Mean monthly altitudinal gradient of temperature (°C/100 m) of Drass and Leh stations with reference to Kargil (2704 m amsl)

| Months | Drass (3066 m amsl) | | Leh (3513 m amsl) | |
|---|---|---|---|---|
| | Max | Min | Max | Min |
| January | − 0.53 | − 2.14 | − 0.17 | − 0.08 |
| February | − 0.21 | − 2.01 | − 0.30 | − 0.03 |
| March | 0.67 | 2.48 | − 0.21 | − 0.12 |
| April | 1.93 | 3.04 | 0.20 | 0.56 |
| May | 2.69 | 3.50 | 0.55 | 0.18 |
| June | 1.41 | 4.64 | 0.44 | 0.28 |
| July | 2.83 | 4.71 | 0.61 | 0.93 |
| August | 3.42 | 4.57 | 0.56 | 0.93 |
| September | 3.36 | 4.25 | 0.49 | 0.87 |
| October | 1.53 | 2.92 | 0.53 | 0.77 |
| November | 1.27 | 2.10 | 0.31 | − 0.65 |
| December | 0.76 | − 2.08 | − 0.04 | − 0.39 |

Leh receives 44.93% of precipitation (rainfall) from the months of June to September followed by Kargil which receives 20.97%, and lowest summer precipitation (rainfall) of 18.63% for the same period is recorded at Drass. The maximum precipitation in the summer season occurred in the month of June at Drass and Kargil, whereas at Leh it is in the month of August. The meteorological observations made during the summer seasons from 2005 to 2009 in the vicinity of accumulation (4910 m amsl) and snout (4160 m amsl) zones of the Durung Drung glacier (Table 3.9) indicate that the highest maximum and lowest minimum temperature in the vicinity of accumulation zone at equilibrium line is 10.1 °C and − 8.3 °C, respectively, whereas the temperature ranged between 18.6 °C and − 4.8 °C near snout, with a mean diurnal variation of 13.5 and 15.6 °C.

### 3.7.2 *Wind*

The wind flows down the slopes of Durung Drung glacier valley from an altitude of 6400 m amsl in southwest to northeast direction at a speed of 5–15 km/h during the summer season. The winds, originating from steep slopes of glaciers, usually move in northeast direction at an average speed of 8 km/h. The winds commencing from Zanskar change their direction from west to east, in the afternoon, and travel at an average speed of 6 km/h. Two types of winds—anabatic and katabatic—flow in the glacier valleys of west Himalaya. The winds when come in contact with heated slopes during the day-time get warmed and expand upward are known anabatic, whereas

**Table 3.6** Precipitation and its parameters at Leh

| Months | Average number of days | %age of days | Average amount of precipitation (mm) | Precipitation as % age of total precipitation/ year | Average amount of precipitation/ day |
|---|---|---|---|---|---|
| January | 1.4 | 10.37 | 11.8 | 10.26 | 8.42 |
| February | 1.1 | 8.14 | 8.6 | 7.47 | 7.8 |
| March | 1.5 | 11.11 | 11.9 | 10.34 | 8.0 |
| April | 0.8 | 5.92 | 6.5 | 5.65 | 8.12 |
| May | 0.7 | 5.18 | 6.5 | 5.65 | 9.28 |
| June | 0.5 | 3.70 | 4.3 | 3.73 | 8.6 |
| July | 1.9 | 14.07 | 15.7 | 13.65 | 8.26 |
| August | 2.3 | 17.03 | 19.5 | 16.95 | 8.47 |
| September | 1.4 | 10.37 | 12.2 | 10.60 | 7.57 |
| October | 0.5 | 3.70 | 7.1 | 6.17 | 14.2 |
| November | 0.4 | 2.96 | 2.9 | 2.52 | 7.25 |
| December | 1.0 | 7.40 | 8.0 | 6.95 | 8.0 |
| | 13.5 | 100% | 115.0 | 100% | 103.97 |

*Source* Indian Meteorological Department

**Table 3.7** Precipitation and its parameters at Kargil

| Months | Average number of days | % age of days | Average amount of precipitation (mm) | Precipitation as % age of total precipitation/ year | Average amount of precipitation/ day |
|---|---|---|---|---|---|
| January | 5.3 | 18.27 | 45.2 | 14.74 | 8.52 |
| February | 4.5 | 15.05 | 51.6 | 16.83 | 11.46 |
| March | 5.5 | 18.96 | 86.3 | 28.15 | 15.96 |
| April | 3.6 | 12.41 | 31.2 | 10.17 | 8.66 |
| May | 2.0 | 6.89 | 23.6 | 7.69 | 11.8 |
| June | 1.3 | 4.48 | 10.3 | 3.36 | 7.92 |
| July | 0.8 | 2.75 | 5.6 | 1.82 | 7.00 |
| August | 1.5 | 5.75 | 10.2 | 3.32 | 6.8 |
| September | 0.8 | 2.75 | 7.6 | 2.47 | 9.5 |
| October | 0.6 | 2.06 | 4.8 | 1.56 | 8.00 |
| November | 0.6 | 2.06 | 4.3 | 1.40 | 7.16 |
| December | 2.5 | 8.62 | 25.8 | 8.41 | 10.32 |
| | 29.0 | 100% | 306.5 | 100% | 113.1 |

*Source* Indian Meteorological Department

**Table 3.8** Precipitation and its parameters at Drass

| Months | Average number of days | % age of days | Average amount of precipitation (mm) | Precipitation as % age of total precipitation/ year | Average amount of precipitation/ day |
|---|---|---|---|---|---|
| January | 8.6 | 15.60 | 126.3 | 16.68 | 14.68 |
| February | 7.9 | 14.33 | 129.9 | 17.16 | 16.44 |
| March | 9.3 | 16.87 | 173.4 | 22.90 | 18.64 |
| April | 7.5 | 13.61 | 100.6 | 13.29 | 13.41 |
| May | 5.2 | 9.43 | 64.0 | 8.45 | 12.30 |
| June | 2.2 | 3.99 | 21.4 | 2.82 | 9.27 |
| July | 1.5 | 2.72 | 13.5 | 1.78 | 9.00 |
| August | 1.7 | 3.08 | 14.6 | 1.92 | 8.58 |
| September | 1.6 | 2.90 | 16.0 | 2.11 | 10.00 |
| October | 2.2 | 3.99 | 17.8 | 2.35 | 8.09 |
| November | 2.0 | 3.62 | 22.9 | 3.02 | 11.45 |
| December | 5.4 | 9.80 | 65.5 | 7.46 | 10.46 |
| | 55.1 | 100% | 756.9 | 100 | 142.77 |

*Source* Indian Meteorological Department

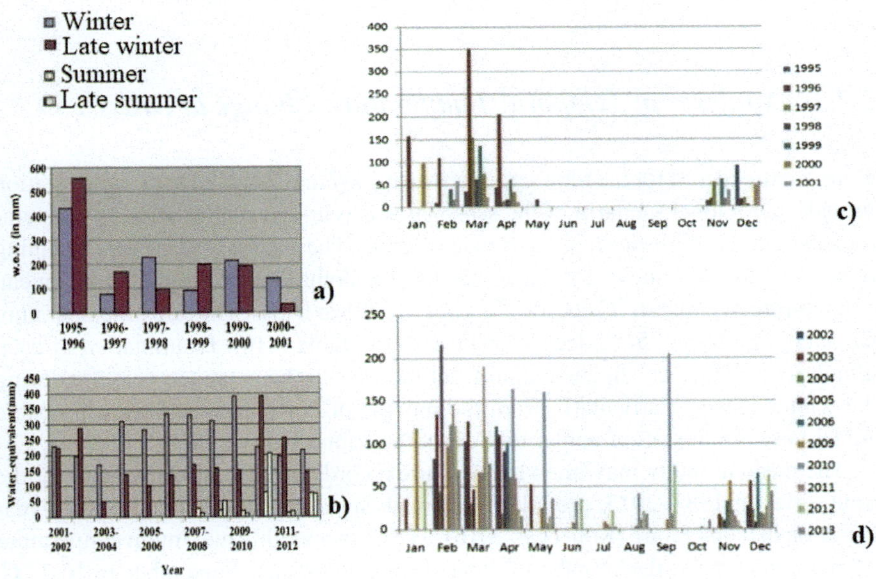


**Fig. 3.5** Graphic representation of **a** winter and **b** late winter precipitation (in w.e.v.) from 1995 to 2001, **c** monthly distribution of precipitation (in w.e.v.) from 1995 to 2001 and **d** from 2002 to 2013 at Drass (Ladakh)

**Table 3.9** Comparative statement of temperature range at snout (4160 m amsl) and Equilibrium Line (4925 m amsl) of Durung Drung glacier

| Snout | | | | | |
|---|---|---|---|---|---|
| Year | 2005 | 2006 | 2007 | 2008 | 2009 |
| | Sept. 22 | July 30 | Aug. 19 | Sept. 17 | Aug. 13 |
| Max. Temp | 13 | 18.2 | 18.6 | 12.6 | 18.4 |
| Min. Temp | − 3.9 | 3.2 | 4.3 | − 4.8 | 3.9 |
| Diurnal Range | 16.9 | 15.0 | 14.3 | 17.4 | 14.5 |
| Mean Range | 15.6 | | | | |
| *Equilibrium line* | | | | | |
| Year | 2005 | 2006 | 2007 | 2008 | 2009 |
| | Sept. 06 | July 24 | July 20 | Sept. 03 | Aug |
| Max. Temp | 5.8 | 10.1 | 9.7 | 4.9 | 7.9 |
| Min. Temp | − 6.5 | − 4.2 | − 3.8 | − 8.3 | − 4.9 |
| Diurnal Range | 12.3 | 14.3 | 13.5 | 14.2 | 12.8 |
| Mean Range | 13.55 | | | | |

katabatic winds are active during night-time when slopes cool down and cool wind sinks down into the valleys.

### 3.7.3 Analyses of Temporal Temperature Change at Drass

Temperature record for the time series between 1987 and 2013 at Drass valley station provides a regional scenario of the seasonal and year-to-year variation in temperature and extent of glacier mass at higher altitude. The record for 26 years spanning from 1987 to 2013 shows that there has been a slight increase in the annual mean temperature by −0.226 °C decade$^{-I}$ earlier to 1995 at Drass. The rate of temperature has increased to 0.375 °C decade$^{-I}$ since 1996 (Table 3.10). Examination of each month record, further, suggests significant increase in the mean temperature during the winter season, particularly from the months of November to May, a probable signal to seasonal shift of winter to the month of June.

The mean monthly maximum temperature record during summer season, for the time series of 1987–2013, provides a regional canvas of seasonal and year-to-year variation in temperature (Koul et al. 2016). The mean monthly maximum temperature during summer for the months of June, July, August and September is 16.2 °C, 20.8 °C, 17.7 °C and 11.7 °C, respectively, for the years 1987–2001, as compared to 20.6 °C, 23.7 °C, 23.6 °C and 20.4 °C for the years 2001–2008 and 14.9 °C, 17.22 °C, 15.17 °C and 10.82 °C for the years 2008–2013. The highest temperature ranged between 26.2 °C and 25.3 °C during the summer season, particularly in the months of July and August, for the time series of 2001–2008 and 1987–2001, respectively,

**Table 3.10** Correlation results of seasonal monthly temperature and precipitation of Drass from 1995 to 2013

| | | Correlation | $p$-value | Significance |
|---|---|---|---|---|
| Winter (December to February) | Mean minimum temperature and precipitation (w.e.v.) from 1995 to 2000 | – 0.18 | 0.772 | No |
| | Mean minimum temperature and precipitation (w.e.v.) from 2000 to 2013 | 0.191 | 0.533 | No |
| Late winter (March to May) | Mean minimum temperature and precipitation (w.e.v.) from 1995 to 2000 | – 0.474 | 0.42 | No |
| | Mean minimum temperature and precipitation (w.e.v.) from 2000 to 2013 | – 0.69 | 0.822 | No |
| Summer (June to August) | Mean maximum temperature and precipitation (w.e.v.) from 2007 to 2013 | – 0.145 | 0.784 | No |

as compared to 12.1 °C and 13.7 °C during the time series of 2008–2013, indicating high range of dispersion in maximum temperature during the time series 1987–2001.

The mean minimum temperature during winter season for the months of November, December, January, February, March, April and May ranged between – 3.635 °C and – 22.45 °C, – 1 °C and – 19.33 °C and – 9.48 °C and 1.02 °C, for the time series 1987–2001, 2001–2008 and 2008–2013, respectively, showing degree variation of standard deviation of 1.98 °C- 4.71 °C, 1.57 °C–4.17 °C and 0.28 °C–1.39 °C. The lowest minimum temperature of -34 °C and -27 °C for winter season is recorded for the months of January and February, respectively, for the time series 1989–2001 and 2001–2008, respectively, as compared to -13.4 °C and – 10.7 °C for the time series of 2008–2013. The monthly diurnal temperature range for winter and summer seasons shows an increase in the trend during the time series 1987–2001 and 2001–2013. The average winter diurnal is 14.9 °C and 11.3 °C for the time series 1987–2001 and 2001–2013, respectively, and summer diurnal is 14.98 °C and 11.576 °C for the time series 1987–2001 and 2001–2013, respectively. The highest diurnal range for the time series 1987- 2001 and 2001–2013 during winter is 18.8 °C and 12.8 °C, respectively, for the month of February and during summer it is 16.48 °C and 12.04 °C, respectively, for the month of July.

### *3.7.4 Precipitation*

The analyses of daily precipitation, from the first fortnight of the month of November of the year 1995 to the last fortnight of the month of October of the year 2001, vary from 984 mm to 171.6 mm in water equivalent volume (w.e.v.) in comparison with 979.9 mm to 219 mm w.e.v. for the time series from 2001 to 2013. The highest (> 500 mm w.e.v.) snowfall of 984 mm was in the year 1995–96, 979.5 mm in the year 2010–11, 565 mm in the year 2007, 562 mm in the year 2008, 589.1 mm in the year 2009 and 587.3 mm in the year 2012. The moderate (between 300 and 450 mm w.e.v.) snowfall is recorded during the time series of years 2001 to 2007 and 2012 to 2014. The low snowfall, < 300 mm w.e.v., is recorded for the time series of the years from 1991 to 1995, 1997 to 1998, 1999 to 2000 and 2000 to 2001, suggesting a dry decade from the year 1991 to 2001 with the year 1995 to 1996 as an exception.

The annual, seasonal and monthly mean snowfall pattern for the time series from the year 1995 to 2001 shows that 196 mm w.e.v. of snowfall took place between the months of November and March of winter season and 206.8 mm w.e.v. of snowfall between the months of March and May of late winter season indicating that about 52% of snowfall took place in the late part of the winter season. Similarly, during the time series from the year 2001 to 2007, 253.4 mm w.e.v. of snowfall took place in the months of November, December, January and February and 155.48 mm w.e.v of snowfall took place in the months from March to May representing late winter season. The snowfall in late winter season amounts to 38% of the total snowfall. In the time series from the year 2008 to 2013, 278 mm w.e.v. of snowfall took place in the winter season, 211.67 mm w.e.v. of snowfall took place in late winter season, and 100.71 mm w.e.v. of precipitation occurred between June and September of the summer season. The late winter rainfall represents 37.5% of the snowfall and precipitation in summer season represents 17%. The Southern Oscillation acts as catalyst to activate Westerlies in the region during summer period causing intense rainfall near snout and snowfall in the vicinity of the equilibrium line as well as accumulation of the glacier. Sometimes, the high-intensity moisture laden winds lead to flash floods in the lower and snowstorms in higher altitudes, as observed during the years 2006, 2007 and 2010 in Durung Drung glacier valley and 2012 and 2014 in Machoi glacier valley. During El Niño-Southern Oscillation (ENSO) year zonal westerly flow becomes quite strong to support the mid-latitudinal system to reach the Indian subcontinent.

## 3.8 Summary

Himalaya is the radiation window of the earth, since thin dry air of higher altitude facilitates the entry of solar radiation by day and rapid loss of heat at night by terrestrial radiation. The rapid day-time heating and night-time cooling results in large diurnal ranges of temperature that is characteristic of the highland climate. At night, during the long winter season of November to May, the terrestrial radiation is rapid and leads to a cooler environment along the ridge slope that induces temperature and precipitation gradient in the winter season which serves as one of the factors favouring the sustenance of the glaciers. The steep temperature gradient during summer season facilitates the cool environment along the ridges favouring the glaciers.

The upper climate zone of western Himalaya comprises Zanskar, Ladakh and Karakoram ranges that receives least precipitation with extremely low density and temperature during winter and remains much below freezing temperature with a result that snow takes much longer time to settle (Khali et al. 2022). The north-west Himalaya is an exceptional orographic chain of the mountain where combined effect of relief and surface configuration plays a significant role in the dynamics of Indian monsoon as well as Western Disturbances. Baspa glacier valley in Dhauladhar (Himachal Pradesh) and Liddar glaciers valley in Pir Panjal (Kashmir) ranges are strongly influenced by Southwest Monsoon during the summer and Westerlies during the winter seasons having positive cascading effect on the health of the glaciers. The trends for four decades from 1980 to 2022 show an annual rate of increase of 0.018 °C $a^{-1}$, 0.032 °C $a^{-1}$ and 0.051 °C $a^{-1}$ in Pir Panjal region, Kashmir valley and Ladakh region, respectively, and general decreasing trend in precipitation with large inter-annual variability (Shafiq et al. 2024).

Differential rates of warming are noticed across Himalaya, such as 0.6 °C/decade over Nepal between 1977 and 2002, 0.2 °C/decade over entire northeast India between 1971 and 2005, 0.16 °C/decade over northwestern Himalayas, 0.23–0.43 °C/decade over western Himalayas and 0.27–0.54 °C/decade over Pakistan Himalayas, exceeding the global average of 0.4 °C (Dimri et al. 2018). The warming in maximum temperature is more prevalent than in the average and minimum temperature. Though overall temperature shows increase in trend in western Himalaya, there are years with either very low or high temperature owing to the complex topography of the western Himalaya. Thus, different thermodynamical and orographic forcings at various locations in the mountain dictate the temperature. Similarly, the normalized precipitation anomaly trends at different stations in the Himalaya are not homogeneous, bringing variability in precipitation amount over the western Himalaya due to different thermodynamical and orographic forcings (Hunt et al. 2018).

Although the teleconnection mechanism between snow anomalies in the Tibetan Plateau and summer monsoon remain largely unknown, heavy snow in the Himalaya during winter or spring has long been regarded as a possible precursor of deficient Indian monsoon rainfall during the subsequent summer (Hunt et al. 2021). The winter snow cover in Himalaya serves as key factor in long-range forecasting of Indian Summer Monsoon (ISM) rainfall, whereas migration of mid-latitude Western Disturbances into the tropics leads to weakening of the ISM (Hunt et al. 2024). The stray El Niño Southern Oscillation also activates the Westerlies during summer that cause intense rainfall near the snout of the glaciers and snowfall at equilibrium line of glaciers. These moisture laden winds cause glaciers to get more snow in winter and pre-monsoon, and thus the glaciers sustain.

The analyses of long time series corroborate the fact that stations located across NW Himalaya are situated at different geographical locations having different altitudes and topographic orientations forming different local weather circulations and hence different temperatures. The analyses suggest that changes in the day-time and night-time temperatures are not symmetrical as warming is more pronounced in the day-time as opposed to the night-time temperatures and the coldest nights are getting warmer and consequently percentage number of cold nights is decreasing while percentage number of warm nights is increasing during winters (December–January–February).

# Chapter 4
# Glacier Geometry and Dynamics

**Abstract** The values of mean rate of mass balance of glaciers in Spiti-Lahaul, Himachal Pradesh (– 0.37 ± 0.09 m w.e.y$^{-1}$), Nepal (– 0.34 ± 0.09 m w.e.y$^{-1}$) and Nyainqêntanglha, Tibet (– 0.62 ± 0.23 m w.e.y$^{-1}$), are in contrast to global mean mass balance of – 0.42 m w.e.y$^{-1}$ showing that glaciers in northwest and central Himalaya are receding at different rates, whereas glaciers in west Kunlun Shan, Karakoram and east Pamir are either in equilibrium or gaining mass. A similar disparity in the mass balance for Ladakh region (Stok glaciers shows – 0.39 ± 0.38 m w.e.y$^{-1}$ during 2015–2019, Pensilungpa glacier shows – 0.36 ± 0.04 m w.e.y$^{-1}$ between 2016 and 2019, Rulung glacier shows – 0.11 m w.e.y$^{-1}$ from 1979 to 1981) and central Himalayan region (Tipra Bank glacier shows – 0.14 m w.e.y$^{-1}$ between 1981 and 1988, Dunagiri glacier shows – 1.04 m w.e.y$^{-1}$ between 1984 and 1990, Chorabari shows – 0.73 m w.e.y$^{-1}$ between 2003 and 2010, Dokriani shows – 0.32 m w.e.y$^{-1}$ between 1993 and 2014) is recorded. Similar trend is noted in the variation of the velocity of west Nepal (– 21.0 ± 2.3% dec), east Nepal (– 17.0 ± 1.0% dec), Bhutan (– 14.5 ± 1.3% dec) compared to Hindu Kush (9.8 ± 2.9% dec), Pamir (– 9.4 ± 1.6% dec) and Tien Shan (– 6.4 ± 1.0% dec) glaciers. The glaciers of Karakoram and west Kunlun with a surface velocity of 3.6 ± 1.2% dec and 4.0 ± 2.1% dec show a significant speed. Besides the global warming, the thickness of debris cover, a non-climatic tool, controls the melting of glaciers. However, 'debris cover anomaly' phenomenon cause no elevation change rates for both debris-covered and clean glaciers in High Mountain Asia. The higher surface area change is associated with glaciers on steep slopes, while no relation is deduced between retreat of terminus and surface slopes. The elevation of ridges, orientation of slope, the amount of winter precipitation and characteristics of summer temperature determine the site, regional pattern and limit of glaciations in Himalaya.

**Keywords** Little Ice Age · Elevation · Aspect · Debris cover · Hypsometry · Hypsometric Integral (HI) · Equilibrium Line Altitude · Slope

Himalaya is geographically an enormously diverse mountain region with a variety of morpho-climatic regimes developed under micro-climate conditions extending from



R. K. Ganjoo and M. N. Koul, *The Himalayan Glaciers, Climate Change and Society*,
https://doi.org/10.1007/978-3-031-78283-1_4

Arunachal Pradesh in the east to Ladakh in the west. It is believed that Himalaya uplifted slowly so that antecedent rivers had no difficulty in continuing to flow through their channel and develop imprints of fluvial markings on the valley slopes of mountain ridges. The gradual uplift of Himalaya led concomitantly to the onset of cold conditions. The low temperature further resulted in developing freezing conditions throughout the year and exceptionally cold environment led to the process of super-saturated permafrost conditions causing hydrostatic pressure along the weak zones of rocks producing polygonal fractures and cracks. The intense cold arid conditions led to the development of a huge amphitheatre along the rind surface of large number of cirques along the protracted slope of higher reaches of Himalaya. Further, uplift of Himalaya pushed the mountain range above the snowline where precipitation is invariably in the form of snow. The Little Ice Age (LIA period), approximately 400 to 700 years ago, experienced a widespread glaciation in Himalaya and the extent of glaciers is considered the benchmark position from which the glaciers are since receding. Studies show that 14,798 Himalayan glaciers have lost ~ 40% of area since the LIA with a loss of ice mass between − 0.011 and − 0.020 m w.e./year (Lee et al. 2021). Besides the role of climate change and topography, the variations in South Asian Monsoon and mid-latitude Westerlies affect the dynamics of the glaciers in Himalaya.

The Indian Summer Monsoon and Western Disturbances, a result of the regional scale interaction between atmospheric and oceanic circulations (Indian Ocean, Atlantic Ocean, Caspian Sea and the associated phenomena), are major tools causing climate variation over Himalaya. The variation in climate gives rise to different glacio-climatic regimes in the Himalaya. Temperature and precipitation gradient is governed by temperature distribution along the mountain slopes in different climatic regimes. The warm type glaciers, where ice temperature is close to melting point throughout its thickness except for the upper accumulation part where temperature is low reaching to freezing conditions, are confined to the narrow valleys of the eastern Himalaya, the central Himalaya and northwest Himalaya (Himachal Pradesh and Kashmir). In the case of warm type glaciers, the melt water exists at the base of glaciers covered with debris. The cold type glaciers, found in the cold desert of Ladakh, are characterized by temperatures below melting point throughout their thickness. The surface and basal melting occurs during the short summer ablation season and supraglacial streams appear temporarily in the lower ablation zones. The cold type glaciers are, by and large, clean and confined to broad valleys.

Most of the high peaks of Himalaya are too steep to retain winter snow on their upper slopes. The glacier formation is severely controlled by topography of mountainous valley where an amphitheatre like depression formed by permafrost activities provides space for the snow to accumulate. The sliding of ice from higher elevation of mountains is a function of glacier thickness and driving stress (horizontal component of the ice weight per unit area). Complexities at the bed/ sole of the glacier (bed roughness, type of bed) constrain the basal velocity represented by a power-law of the driving stress (Dehecq et al. 2018). The change in the thickness of glacier and slope together accounts for the change in driving stress.

Besides the accumulation of snow, avalanching, basal freezing and growth of superimposed ice undergo a series of changes which transform the snowfall into glacier ice. The transformation of snow survives the summer melt season, to form loosely consolidated and randomly oriented ice crystals called firn, with interconnecting air passage having density > 0.4 $mg/m^3$. Further transformation of firn involves regrowth of ice crystals and elimination of air in passage. Consolidation proceeds sufficiently to isolate the air into separate bubbles; the firn becomes ice with densities between 0.8 $mg/m^3$ and 0.85 $mg/m^3$. Besides snowfall and hail, wind drift snow, avalanches, growth of superimposed ice, freezing of water and condensation of vapour contribute to the mass of the glacier. Snow accumulated over the glaciated surface is first to melt as a result of high temperature during the melting season that prevails from March to May. The debris over the ice surface, mixed with extraneous matter, decreases albedo appreciably to increase rate of melt. The debris-land and debris-lake glaciers combined contributed 47% (south) and 46% (north) to total volume loss of Himalayan glaciers ($n = 14{,}798$), despite comprising only 6% and 9% of the number of glaciers, respectively (Lee et al. 2021), signifying the strong control of presence/absence of a debris-cover.

The changes in glaciers are also affected by changing temperature and precipitation pattern across different sub-regions of Himalaya from east to west that decides the variable timing of deglaciation (Khadka et al. 2023). Other meteorological parameters, e.g. wind, moisture, heat from the bedrocks and, to some extent, rain falling on the glacier surface, also contribute to the change in glaciers. The melting glaciers contribute to the major rivers emanating from Himalayan region assuring water security and meeting challenges of scarcity in the lowland areas.

The Himalayan glaciers are regarded as storehouse of large glaciers, outside the polar region in comparison with Alpine glaciers, which are under great stress because of climate change phenomena and may contribute to rise in sea level. The warming of oceans causes rise in the sea surface temperature (SST) and heat content which in turn leads to the thermal expansion and rise in sea level. The rate of warming of tropical Indian Ocean is fastest with rise in North Indian Ocean (NIO) at the rate of 3.3 mm/year between 1993 and 2017 (Roxy et al. 2020). The rural population in Low Elevation Coastal Zones (LECZ) is most vulnerable to the rising sea level due to receding coastlines compounded by land subsidence at an average rate between ~ 10 mm/year and 32 mm/year (Horst et al. 2018; Haque et al. 2019). The extremely vulnerable population of LECZ (the Center for International Earth Science Information Network defines 'Low Elevation Coastal Zone' as a land area with elevation of up to 20 m) increased from 11.9 to 13.4% in Thailand, declined from 5.5 to 4.3% in Sri Lanka whereas in Myanmar, Bangladesh and India the rural population declined but urban LECZ population increased between 1990 and 2010 (MacManus et al. 2021 and references herein).

Changes in temperature, snowfall and snow extent, reported from Himalaya conflicts with change in climate. It is evident from the studies in northwest and central Himalaya that glaciers are receding at different rates (Das and Sharma 2019). In contrast to the global mean mass balance, since the turn of twenty-first century, of – 0.42 m w.e.$y^{-1}$, the mean rate of mass balance loss of glaciers in Spiti-Lahaul is

$-0.37 \pm 0.09$ m w.e.y$^{-1}$, west Nepal is $-0.34 \pm 0.09$ m w.e.y$^{-1}$, Nyainqêntanglha is $-0.62 \pm 0.23$ m w.e.y$^{-1}$, whereas the glaciers in west Kunlun Shan, Karakoram and east Pamir are either in equilibrium or gaining mass (Dehecq et al. 2018 and references herein). Similar trend of variable mass balance loss is observed for the glaciers of northwest and central Himalayan region. In Ladakh region, Stok glaciers show a mass balance loss of $-0.39 \pm 0.38$ m w.e.y$^{-1}$ during 2015–2019 (Soheb et al. 2020), Pensilungpa glacier in Suru valley shows a loss of $-0.36 \pm 0.04$ m w.e.y$^{-1}$ between 2016 and 2019 (Mehta et al. 2021), Rulung glacier shows a loss of $-0.11$ m w.e.y$^{-1}$ from 1979 to 1981 (Srivastava et al. 1999), whereas in central Himalayan region the Tipra Bank glacier shows a loss $-0.14$ m w.e.y$^{-1}$ between 1981 and 1988, Dunagiri shows a loss of $-1.04$ m w.e.y$^{-1}$ between 1984 and 1990, Chorabari shows a loss of $-0.73$ m w.e.y$^{-1}$ between 2003 and 2010, and Dokriani shows loss of $-0.32$ m w.e.y$^{-1}$ between 1993 and 2014 (Garg et al. 2022b). Interestingly, surface velocity of Himalayan glaciers shows a decrease towards the east. The velocity of glaciers in west Nepal is $-21.0 \pm 2.3\%$ dec, east Nepal is $-17.0 \pm 1.0\%$ dec, and Bhutan is $-14.5 \pm 1.3\%$ dec as compared to the surface velocity of $-9.8 \pm 2.9\%$ dec, $-9.4 \pm 1.6\%$ dec and $-6.4 \pm 1.0\%$ dec of Hindu Kush, Pamir and Tien Shan glaciers, respectively (Neckel et al. 2017). The glaciers of Karakoram and west Kunlun with a surface velocity of $3.6 \pm 1.2\%$ dec and $4.0 \pm 2.1\%$ dec show a significant speed.

The research on glaciology, during last two decades, has developed to elucidate the current behaviour of glaciers of Himalaya and trans-Himalaya specially with a focus on the volumetric changes based on GIS and remote sensing techniques. The application of remote sensing and GIS has been useful in mapping the glacier margins, estimating glacier area, deciphering glacier length, hypsometry, mass balance estimates, etc. Orientation of multi-temporal Landsat and IRS LISS III images of different time series applying standard band-ratio technique and on-screen delineation has provided glacier-wise details with respect to glacier identification number, glacier location, measurement of dimensions, etc. The technique has been helpful in comparing the DEMs of different time series and calculating thickness variation using algebraic software.

## 4.1 Natural Determinants of Glaciers

The elevation of ridges, orientation of slope, the amount of winter precipitation and characteristics of summer temperature determine the site, regional pattern and limit of glaciations in Himalaya. The natural determinants dictating the glaciation in Himalaya are explained further.

### 4.1.1 Elevation

The Himalaya-Tibetan Orogen, due to its extreme high altitude (6500 m amsl) and large expanse acts as a strong orographic barrier to maritime winds to influence the global circulation particularly south Asian Monsoon (Spicer 2017; Ding et al. 2017). The maritime winds bring moisture to high peaks of Himalaya and trans-Himalaya, playing a vital role in temporal distribution of glaciers, in both the past and present. The present-day climate of Himalaya-Tibetan massif, dominated by two atmospheric circulation systems, the mid-latitude Westerlies and Indian Summer Monsoon, results in recharging the glaciers, such as Drass, Suru, Zanskar, Nubra and Shyok, in Himalaya and trans-Himalaya, particularly in the Upper Indus basin.

The northwest Himalayan glaciers have their origin and termination at critical elevations of 3700 m amsl and 5000 m amsl and as such a large area of Himalaya falls above the snowline that serves a primary factor in deciding the snow extent and condition of glaciers in northwest Himalaya. The rise in the elevation of Himalaya towards west to towering heights from 6000 m amsl to 7500 m amsl supports the long survival of glaciers. The uplift of Pir Panjal and Dhauladhar ranges during Pleistocene period continued through Holocene which led to the maximum extension of ice cover in the central Himalaya between 2600 m amsl and 2800 m amsl and between 2000 m amsl and 2250 m amsl in the western Himalaya in winters. The terminal moraines in the higher Himalaya and Karakoram are recorded at an elevation of 3800 m amsl. The mean elevation of glaciation varies from 6000 m amsl in Ladakh and Kashmir in the northwest to 4600 m amsl in Arunachal Pradesh in the northeast.

Glaciers of Himalayan region, in general, show loss of mass and shrinkages since Little Ice Age around 1850 AD (Singh et al. 2021b), whereas glaciers occupying the west end of Himalayan arc (Karakoram and adjoining High Himalaya) show positive mass area change and the Karakoram glaciers are either advancing or stable (Krishnan et al. 2019). Overall mass balance of Karakoram glaciers from 2000 to 2016 was $-0.03 \pm 0.07$ m w.e. $y^{-1}$ showing nearly a stable mass balance of $-0.020 \pm 0.064$ m w.e. $y^{-1}$ in the western and $-0.101 \pm 0.058$ m w.e. $y^{-1}$ in the eastern parts of Karakoram (Lhakpa et al. 2022 and references herein). Further, the glaciers in the eastern part of Karakoram showed glacier elevation change of $-0.19 \pm 0.22$ m $y^{-1}$ compared to surge ($-0.18 \pm 0.22$ m $y^{-1}$) and non-surge ($0.20 \pm 0.22$ m $y^{-1}$) type glaciers in western part.

Glaciological studies on elevation changes and mass balance on selected glaciers of Himalaya indicate that the glaciers located at higher elevations have deglaciated (melted) at lower rate compared to the glaciers located at the lower elevations. The studies explain the factors of debris thickness and corresponding insulating effect with increasing elevation to the variation in elevation change. Hence, transition zone of debris covered, and clean ice is where thickness of debris is likely to be least and the elevation change is mostly negative (Bhushan et al. 2018 and references herein). It is further noted that glaciers with debris limited to lower tongues experience minimum surface lowering at higher elevations compared to the glaciers where trunk is extensively covered with debris and located at limited toe distance / elevation from the terminus. Significant negative correlation ($r = -0.5$) between minimum elevation and terminus change is noticed, while maximum elevation exhibits significant positive relation ($r = 0.4$) indicating that high frontal change is dominant in the glaciers located in low altitudes (Das and Sharma 2019), a trend observed for the large glaciers too (Garg et al. 2017). Earlier studies on western and central Himalayan glaciers also revealed similar relationship between elevation range and change in glacier front (Bhambri et al. 2020).

In a highly dynamic terrain of Himalaya, glacier morphology (e.g. shape, size and hypsometry), steep slope and large accumulation area are the drivers to decide for the movement of ice to lower elevation and add more mass to the lower ablation zone of the glaciers, bearing the fact in mind that large valley glaciers have long response time (Das and Sharma 2019).

## *4.1.2 Aspect*

Aspects regulate the amount of direct solar energy distribution, direction of wind circulation (Buri and Pellicciotti 2018), and control the distribution of snowfall also. The glaciers facing north directions (N, NE and NW), such as the ones in the Liddar valley (Pir Panjal) and Baspa valley (Dhauladhar), recede at a slightly low rate than the glaciers facing south directions (S, SE and SW). The large (66.8%) and long-sized (18.4 km) glaciers, facing the north direction, receive less amount of solar radiation and consequently recedes relatively slowly. The south facing glaciers in Chenab basin show reduction of 1.71% in SE facing, 2.06% in S and 2.57% in SW facing glaciers, whereas the east facing glaciers showed no reduction from 2000 to 2019 (Sharma et al. 2022a). The studies further reveal that the N- and S-oriented glaciers in Ravi basin indicate marginal increase in areas by 0.13% and 0.08%, respectively, from 2000 to 2019, whereas no change is noticed in the E-oriented glaciers. Marginal reduction in area of ~0.81%, 1.29% and 1.34% is noted in the glaciers oriented towards the SE, SW and W, respectively, for the same period. The glaciers in Sutlej basin bearing a N aspect show highest loss of 1.14% in two decades (2000–2019). Sharma et al. (2022a) in the studies highlighted highest loss of 1.84% in SE-oriented glacier in Bhagirathi basin for the corresponding period. Studies by authors suggest a fair percentage (24%) of the glaciers-oriented to E, followed by glaciers oriented to NW (16%) and W (16%), directions in Nubra valley show gain in the snow cover area between 1989 and 2001, whereas 23% of glaciers oriented to SE, 19% oriented to NE, 15% oriented to each S and E directions show loss in the snow cover for the corresponding time. The studies further show that 49% glaciers oriented to N, 25% to SW, 13% oriented to each W and SE gain the snow cover area from 1969 to 1989, and subsequently, there is loss of area between 1989 and 2001.

### 4.1.3 Debris Cover

The debris-covered glaciers are also called as dirty glaciers where a part of ice of ablation zone is covered with apron of rock fragments along with loose assorted material carried from valley walls and deposited on the glacier surface by the action of avalanches or mass wasting processes. Generally, the debris is derived from surrounding walls of the glacier valley (headwalls, sidewalls, basal erosion and moraines) because of weathering and denudation due to the freezing and thaw process. The debris covers the frontal part of the ablation zone of clean glacier ice and is one of the most significant non-climate drivers in the melting of glaciers. The thickness of debris cover controls the response of glaciers to climate change, melting of glacier will enhance under the cover of thin debris and thick cover of debris will insulate underlying ice and suppress melting. However, recent studies found similar elevation change rates for both debris-covered and clean glaciers in High Mountain Asia, referred to as 'debris cover anomaly' phenomenon (Rounce et al. 2021 and references herein). The ice flow plays primary role in governing the advective debris transport from the accumulation to ablation zone of the glacier (Wirbel et al. 2018). The complex interaction between climate, debris load, ice dynamics, hydrologic conditions and topography governs the dynamics of debris-covered glaciers (Huo et al. 2021 and references herein).

The extreme glacial aridity and desertification have led to the formation of thick debris cover throughout the Hindu Kush Himalaya region that distinguishes the region from rest of the globe. The dynamics of glaciers covered with debris differs from that of the clean-ice glaciers due to the insulating effects caused by debris material. The retreat of the front of glaciers is subject to glacier surface morphology, such as thickness of debris cover or supraglacial material and formation of ponds or lakes (Buri and Pellalongicciotti 2018; Shukla and Garg 2019; Huo et al. 2021). The assessment of retreat of glacier fronts, applying remote sensing techniques, under the conditions of different surface morphology may assist in ascertaining the status of glaciers in relation to the oscillation in climate of Himalaya (Bhambri et al. 2020).

According to a study (Herreid and Pellicciotti 2020), about 44% of the world's glaciers are covered with debris measuring > 0.1 $km^2$, 15% have a prominent cover of debris measuring > 1.0 $km^2$ and >7% of the surface area of glaciers measuring > 2 $km^2$ forming 20% world's glaciers are covered with debris. A debris layer that is less than a few centimetres thick can enhance the local melt rate, while a debris layer with a thickness greater than a few centimetres causes an exponential decrease in the local melt rate, scaling with increasing debris thickness (Herreid 2021 and references herein). Increase in the extent and thickness of debris cover on Himalayan glacier surface due to loss of mass, over a period, has slowed the rate of recession as compared to the clean glaciers (Das and Sharma 2019). The field studies, further, corroborate to the fact that glacier surface areas covered with thick debris when exposed to short period of ablation season have a reduced rate of melting (ablation rate) in comparison with glaciers covered with thin debris (Rowan et al. 2021), implying that the temperature reaching the surface of glaciers decreases in proportion to the increase in the thickness of debris.

Study of glaciers in western Ladakh has emphasized significance of supraglacial debris cover, a non-climatic factor, on the glacier dynamics with respect to the modification of surface ablation and reduction in rate of melting (Mandal et al. 2024). The rate of melting of glacier surface initially increases with the thickness of debris cover but decreases after the critical thickness (the thickness of the debris layer at which the ablation rate of a debris-covered glacier is same as for a debris free glacier) is achieved. The retreat of the glacier fronts gets retarded due to the insulating effect caused by the debris cover and waste is down more in the zone of thick debris-covered ice part of glacier (Patel et al. 2021a). Consequently, the glacier fronts are near stable in some of the glaciers of northwest and west Himalaya, whereas in some cases (such as Pensilungpa glacier in Suru valley, Ladakh) high rate of melting is observed near dirty and thin debris-covered ice surface from 2000 to 2020 (Romshoo et al. 2024a) compared to the other glaciers, Haskira, Kange, Hagshu and Durung Drung, in same region and climate zone. Studies showed that debris thickness of Chandra basin glaciers (Bara Shingri, Batal and Kunzum) varied from 0.5 cm to 326 cm with thicker debris cover over the glaciers occupying lower elevations and thin cover over the glaciers occupying higher elevations (Patel et al. 2021a). Studies further suggested that the warming of debris surface during ablation period (June to September) controlled the debris surface temperature which shows decreasing trend with gain in altitude. A strong correlation is noted between the debris surface temperature and variability in air temperature, but wherever low the local weather, topography and aspect conditions interfere with the correlation.

Study of glaciers, in Warwan and Bhut sub-basins of Chenab valley (western Himalaya), using satellite images of 2001 to 2011 indicates variation in debris cover up to 10% in case of the glaciers covering an area up to 10 $km^2$, whereas glaciers covering an area up to 30–40 $km^2$ show variation up to 23% (Space Applications Centre 2016). The study further shows a loss of ~ 3.4% in glacial area for debris-free small glaciers, whereas the loss is 1.2–1.5% for debris-covered small glaciers. The large glaciers with a debris cover of 12–25% show negligible retreat compared to debris-free large glaciers showing a loss of ~ 0.5% in glacial area for the period from 2001 to 2011.

The glaciers with debris cover category of 1–25% in Alaknanda basin showed highest cumulative area loss of 0.40%, whereas in the category of 51–75% the loss is 0.01% during 2000–2019 (Sharma et al. 2022a). Similar trend is observed for the glaciers of Bhagirath basin. The glaciers with debris cover of 76–100% showed an overall positive change in the cumulative glacierized area with an addition of 0.06 $km^2$ between 2000 and 2016, which subsequently increased to 0.16 $km^2$ till 2019, whereas glaciers with debris cover of 51–75% registered an increase of 0.02 $km^2$ in total area from 2016 to 2019 (Sharma et al. 2022a). It is corroborated on the basis of study of Survey of India topographical sheets that in Alaknanda sub-basin more retreat (14%) of glaciers is observed than of Bhagirathi sub-basin (11%) and the retreat is 10% and 1.8% based on the study of satellite images for the sub-basins, respectively (Space Applications Centre 2016).

### 4.1.4 Hypsometry

Hypsometry is a hypsographic curve or cumulative distribution function of elevation in a geographic area. The relation between hypsometry and sensitivity of mass-balance change at Equilibrium Line Altitude (ELA) of a glacier can be used reliably to estimate the sensitivity of changes in ELA of unmeasured glaciers. Hypsometry (area-altitude relationship) of glaciers plays a critical role in the response of terminus ELA or changes in mass balance of glaciers. The ELA is used as a base level to measure the elevation and depression of glacier profile, particularly about accumulation and ablation area and serves as a first-order estimator of mass balance sensitivity. Based on the Hypsometric Integral (HI) calculation, the fronts of top-heavy glaciers are resistant to retreat compared to bottom-heavy glaciers. However, glaciers in the Baspa valley having equidimensional hypsometric distribution are prone to the retreat of terminus compared to top-heavy glaciers. The Durung Drung glacier with bottom-heavy geometry receded at a rate of $-5.05$ my$^{-1}$ during the period 2009–2011.This is because of the accumulation of more mass in the ablation zone as a result of avalanches that enhanced the downward movement of bottom-heavy glaciers due to gravity, e.g. Machoi, Kangriz. Hypsometrically, it can be inferred that the Drang Drung (Durung Drung) glacier is more susceptible to changes in ELA compared to Pensilungpa glacier (Jasrotia et al. 2023). The large glaciers (exceeding the size of 10 km$^2$) placed within the elevations of 5030–5367 m asl in Beas basin show no difference in ELA between 2000 and 2019, whereas glaciers situated above 5129 m elevation in Bhagirathi basin and exceeding the size of 10 km$^2$ showed pronounced changes in the ELA from 2000 to 2016 with a maximum contraction of 21 m in the ELA while most glaciers reflected lowering of the ELA with a maximum magnitude of 13 m (Sharma et al. 2022). A similar trend is noticed for the glaciers, exceeding 10 km$^2$ and placed above 4814 m amsl, in Alaknanda basin where most of the glaciers showed retreat in ELA.

The topography and hypsometry vary with the change in elevation and the effects of seasonal parameters, such as temperature and precipitation, become pronounced. The hypsometric profile plays a significant role in seasonal cloud accumulation and the subsequent precipitation in western Himalaya, situated in the monsoon transition zone, receives both summer monsoonal precipitation and substantial winter precipitation from the Westerlies (Sam et al. 2018). The combined HI for western Himalayan glaciers is 1.13, which is at the boundary between equidimensional and bottom-heavy profiles.

---

HI = Maximum elevation - Mean elevation / Mean elevation - Minimum elevation
Based on the thresholds, five categories of glaciers exist - very top-heavy (HI $< -1.5$), top-heavy ($-1.5 <$ HI $< -1.2$), equidimensional ($-1.2 <$ HI $< 1.2$), bottom-heavy ($1.2 <$ HI $<1.5$) and very bottom-heavy (HI $> 1.5$) (Jiskoot et al.)

---

In the states/ union territories of Uttarakhand, Sikkim and Jammu and Kashmir, the hypsometry shows that the elevation changes after approximately 5000 m amsl tend towards a positive value, and glaciers in Himachal Pradesh, Bhutan and Arunachal

Pradesh show a negative elevation change regardless of fact that accumulation mostly occurs at higher elevation. The glaciers in Nepal show no change in elevation after ~ 6800 m amsl (Bandyopadhyay et al. 2019).

### 4.1.5 *Slope*

The degree and orientation of slope playa dominant role in the growth / retreat of glaciers. It is partly due to drifting snow into high accumulation basin by wind action, but more because of the difference in the incoming radiation between north and south facing slopes. Slopes, a crucial topographic component, are a significant controlling factor in determining the advance or retreat of glaciers in the background of climate change. Faster rate of retreat and deglaciation is noticed in glaciers with steep slopes compared to glaciers with gentler slopes.

Glaciers with steep slopes tend to show higher surface area change, while retreat of terminus shows negative relation ($r = -0.5$) with surface slope. In Suru-Zanskar valley, it has been observed that glaciers with steep slopes (> 25°) experienced low rate (– 2.31 $my^{-1}$) of retreat of terminus, while the glaciers with < 25° slope experienced higher rate (– 5.56 $my^{-1}$) of retreat. Studies reveal that Drang Drung glacier covers larger area with higher slopes in accumulation zone and Pensilungpa glacier covers larger area with lesser slopes in the ablation zone, with a consequence that rate of loss of length and area is faster in Drang Drung glacier compared to Pensilungpa (Jasrotia et al. 2023). The large-sized and low altitude glaciers in Jankar Chu watershed, Lahaul, have experienced a high rate of retreat compared to the small-sized glaciers located in high altitudes (Das 2022). The steep slopes are the source for abundant debris on the lower part of the Patsio glacier (western Himalaya) and shading effect of surrounding narrow valley acts as an insulator and reduces the melting of ice (Angchuk et al. 2021 and references herein). A change of 5.92% in glacier cover, between 2000 and 2019, is recorded in the glaciers with mean slope angle of $30^0$ to $35^0$ compared to change of 1.29% in glaciers with slope angle between $15^0$ and $20^0$ in Yamuna watershed (Sharma et al. 2022a). Further, glaciers with gentler slopes of 15°-20° register increase of 1.98% in the glacier area between 2000 and 2019, and glaciers with slope angles between 20° and 25° record maximum loss of 3.33% in Beas catchment. The studies further reveal a reduction of 0.85% in the area of glaciers with slope angle $10^0$–$15^0$ and reduction of 1.89% in the glaciers with slope angle $30^0$–$35^0$ during 2000–2019 in Ravi basin.

## 4.2 Pattern and Extent of Existing Glaciers

Jasrotia et al. (2023) registered a loss of 8.16 $km^2$ accounting for 1.73% in the area of Drang Drung glacier from 1976 to 2020 and 2.25 $km^2$ accounting for 9.84% in the case of Pensilungpa glacier in Zanskar and Suru valley, respectively, demonstrating higher loss of area in Drang Drung. The retreat in both glaciers showed fluctuations. The maximum retreat of Drang Drung glacier was 325.39 m between 2010 and 2015 whereas minimum was 65 m between 2015 and 2020. On the other hand, retreat of 109.26 m occurred between 2000 and 2010 and 45.50 m between 2015 and 2020 in Pensilungpa glacier.

The areas of glaciers in Koshi, Gandaki, Karnali and Mahakali basins in Nepal showed a heterogeneous loss of ~ 24% between 1980 and 2010 (Khadka et al. 2023 and reference herein). The area of glaciers in Yamuna basin, in general, register reduction of 2.13% between 2000 and 2019 with only two glaciers exceeding 10 $km^2$ in sizes showing the least (1.05%) in 2000–2019 and small-sized glaciers showing higher loss of 4.06% (Ajai 2018). The glaciers in Chenb basin ($n = 29$) grouped in area category of 5.01–10.00 $km^2$ show reduction of 3.06% in area from 2000 to 2019, while glaciers falling in the size class of 1.01–5.00 $km^2$ reveal a reduction of 1.88% in cover area and glaciers falling in size class of 10 $km^2$ show limited reduction of only 0.13% in area in two decades (2000–2019) (Sharma et al. 2022a). The studies further suggest a maximum loss of glacier area of 2.36% in glaciers size category 0.51–1.00 $km^2$ and least loss in glaciers of size category of less than 0.50 $km^2$ in Bhagirathi basin. The glaciers in the size category of < 0.50 $km^2$ registered highest percentage of area loss (0.48%) in 2000–2019 in Alaknanda basin. The Upper Teesta watershed (Sikkim) showed lowest rate of retreat of cumulative glacier area in 1988–2001, due to cooler late summer temperatures along with ample rainfall to arrest the deglaciation (Debnath et al. 2019). Highest rate of change in glacierized area is noticed in 2001–2016, with a loss rate of 0.66 $km^2$ per annum (Sharma et al. 2022a). There is no significant change in the area of Siachen glacier since 1980 but for moderate thickening at high altitude and significant thinning in lower altitude (Agarwal et al. 2017). A recent study in Chandra basin (Lahaul and Spiti region) identifies fast rate of deglaciation (20–30%) between 2001 and 2011 compared to the rate (10–20%) during 1980–2001 (Dhar et al. 2020).

The maximum advance of glaciers in western Himalaya is documented between > 60 ka and 85 ka (late Pleistocene period) when the thickness of ice remained over 250 m from the present valley floor and Gangotri glacier terminated at Jhala (2400 m amsl), about 40 km downstream of present Gaumukh (4000 m amsl) (Sharma et al. 2022a).

The glaciers in Himalaya are located in the remote and treacherous terrain that is poorly connected and inaccessible with all-weather motorable roads. Monitoring glaciers by direct field method, therefore, is difficult whereas using remote sensing and GIS technique provides an advantage of giving a synoptic view of the glaciated region on a regular basis. The updated glacier inventory of the northwest and west Himalaya with demarcation of the glacier boundaries situated in Liddar valley, Pir Panjal ranges, Baspa valley, Dhauladhar ranges, Sutlej basin, Drass valley, Suru-Zanskar valley of Upper Indus basin and Nubra-Shyok valley of Ladakh is worked out on the basis of combined study of Survey of India (SOI) topographic sheets, aerial photographs, Landsat and IRS LISS III images of the years 1965, 2001 and 2013, supplemented with ground verification, wherever possible. The authors carried out extensive field investigations of glaciers with a thrust to verify and demarcate the snouts accurately and also assess changes in glacier area, particularly in the case of debris-covered glaciers. The methodology adopted involved the documentation and mapping of snouts with the help of DGPS and Set 2C survey station to delineate margins of active ice cover, which are otherwise difficult to discern in satellite images. The field data has been digitized in collaboration with the scientists at Space Applications Centre (SAC, Ahmadabad (India), and cross-verified with satellite images of time series.

The base map of the area of study was prepared from the Survey of India (SOI) topographic maps of 1965 of the scale 1:50,000. Only seven SOI topographic sheets of Ladakh region including Nubra and Drass-Suru-Zanskar sub-basin were available that covered 115 glaciers of Drass, 268 glaciers of Suru-Zanskar and 114 glaciers of Nubra-Shyok. Out of the 115 glaciers of Drass, 81 glaciers covering > 0.1 $km^2$ area were selected for detail investigations. Using Geographic Information System (GIS) techniques, the boundaries and area of glaciers were delineated after the satellite images were geo-referenced using SOI topographic maps. Besides, various other features such as accumulation area, ablation area, equilibrium line and depositional feature like moraines of glaciers were delineated more accurately by using GIS False colour satellite images and field verifications.

## 4.3 Northwest Himalayan Glaciers

### *4.3.1 Liddar Valley Glaciers: Kashmir Himalaya*

The Liddar valley occupies the southeast part of giant Kashmir Himalayan synclinorium and forms part of the trans-Himalaya. It lies between the Pir Panjal range in the south and southeast, the north Kashmir range in northeast is the main Himalayan range and Zanskar range is in northwest. The valley is girdled on three sides by lofty ridges, the Saribal-Katsal ridge (4800 m amsl) in the east, the Wokhbal (4200 m amsl) in the west and Basmai-Kazimpathlinbal (4800 m amsl) in the north. Liddar valley begins from the base of the two main snow fields, the Kolahoi (west Liddar) and the Shishram (east Liddar). The aerial-photographic data and Survey of India topographic maps of 1962 have been used in mapping the margins of glaciers to evaluate the change in the area, length and relative relief of the glaciers from 1962 to 1985, corroborated by extensive field verifications for the period from 1983 to1985. Out of total 37 glaciers, 19 glaciers having > 0.1 $km^2$ area were selected for detailed comparative study. Three major glaciers—Kolahoi (11.9 $km^2$), Shishram (9.9 $km^2$) and Sonsar (6.0 $km^2$)—are confined between an altitude of 3690 m amsl and 5000 m amsl and were extensively surveyed and studied in the field (Tables 4.1, 4.2 and 4.3). The present distribution of glaciers is notably asymmetric. Most of them flow towards north and northwest. The north-flowing glaciers are between 2 and 6 km, and the south-flowing ones are < 2 km in length. Most of the glaciers occupy cirque basins confined along the precipitous slopes of the mountain ranges. Kolahoi, one of the longest glaciers (35 km) in Kashmir valley, extended up to Pahalgam during the Pleistocene time. It is observed that glaciers in Kashmir valley have lost 28.89 ± 0.1% of the area and 19.65 ± 0.069% of volume with variable inter-decadal rates of recession during last 51 years from 1962 to 2013 (Marazi and Romshoo 2018). The area of glaciers shrunk from 12.2 $km^2$ in 1962 to 11.61 $km^2$ in 2010. The Energy and Resources Institute (TERI, India) installed an Automatic Weather Station at Satlanjan (3925 m amsl) equipped with weather sensors to conduct a detail meteorological study and to measure the mass balance of Kolahoi glacier. The team recorded water level at Liddarwat discharge station and measured the snow density at accumulation zone of the glacier. Besides, measurements included assessing the changes in glaciers from 1980 to 2015 using satellite images. The analyses revealed reduction in the length of glaciers by 10%, in glacier boundary (aerial surface) by 13.5% and in volume by 18%. The glaciers vacated about 1% of its length from 1980 to 1990 and 1.1% from 1990 to 2000. However, in next decade, the rate of retreat spiked to 5.4%. The recent studies reveal that Kolahoi glacier has retreated by 2.85 km in past 157 years from 1857 to 2014, with an average retreat of about 18.2 $my^{-1}$ including the retreat of 2.18 $km^2$ and loss of 0.30 $km^3$ of ice volume in last 51 years from 1962 to 2013. The deglaciation between 1990 and 2000 led to the fragmentation of the glacier into four and enhanced the rate of retreat to 0.76%$y^{-1}$ from 2000 to 2018. The rate of retreat of fragments of glacier increased from 24.4 to 27 m between 2013 and 2018 (Rashid et al. 2017, 2020; Shukla et al. 2017). Lone et al. (2018) suggested reduction in the Kolahoi glacier area from 14.46 ± 0.34 $km^2$in 1962 to 11.28 ± 0.21 $km^2$ in 2014, thereby showing disparity in the rate of retreat and deglaciation.

### *4.3.2 Baspa Valley Glaciers: Kinnaur Himalaya*

Baspa valley is situated in Sangla Tehsil of Himachal Pradesh. The valley, NW–SE oriented, is 72 km long and 15 km wide. The valley is girdled on three sides by high Himalaya that houses 89 glaciers. The valley commences from the main ice field situated in southeast from where two main glaciers, Arsamnag (6159 m amsl) and Baspa Bamak (5760 m amsl), commence and are oriented to west. The valley is fenced by Whoming (5215 m amsl), Khomsingh (5746 m amsl) and Khimloga (5712 m amsl) peaks in south, that also forms a water divide between Sutlej and Ganga drainage basins. The valley has an area of 1100 $km^2$ out of which 22.4% is covered by glaciers. The major glaciers, Hania, Janapa, Jora, Saro, Shusha and Naradu, are confined between an altitude of 6000 m amsl and 4500 m amsl. Also, large number of small-sized glaciers occupy cirques and niches. Most of the glaciers (79.4%) are north facing and range in length between 5.5 and 17 km. The equilibrium line altitude (ELA) of present-day glaciers lies between 5200 and 5000 m amsl. The glaciers, at present, are relics of the mighty predecessors that occupied the main valley at an elevation as low as 2500 m amsl and trunk valley at an elevation of 3400 m amsl. The detailed analysis of high-resolution satellite data identified 109 compound valley glaciers with an area of 187 ± 3.7 $km^2$ that were mapped and

**Table 4.1** Surface ablation estimates of Kolahoi and Shishram glaciers on the basis 0s stake measurements

| No. of stakes | Date | Initial length of stake (in cms) | Dat e | Adjusted length of stake (in cms) | Initial length of stake (in cms) | Date | Adjusted length of stake (in cms) | Ablation rate (in cms) | GL ACI ER |
|---|---|---|---|---|---|---|---|---|---|
| 2 | 4th Sept | 143.7 | 7th Sep t | 156.7 | Down | 10th Sept | Down | 13.0 (4 -7 Sept) | KO LA HOI |
| 9 | | 132.8 | | 142.8 | 140.5 | | 170.0 | 39.5 (4–10 Sept) | |
| 8 | | 125.4 | | 133.7 | 131.8 | | 158.3 | 34.8 (4–10 Sept) | |
| 6 | | 153.0 | | 170.4 | Down | | Down | 16.7 (4–7 Sept) | |
| 4 | | Stakes lost | | Stakes lost | Stakes lost | | Stakes lost | | |
| 3 | | 140.6 | | 151.2 | 148.6 | | 179.2 | 41.2 (4–10 Sept) | |
| 7 | | 137.2 | | Down | 140.2 | | 148.3 | 8.1 (7–10 Sept) | |
| 1 | | 119.7 | | 134.1 | 133.4 | | Down | 14.4 (4–7 Sept) | |
| 5 | | 122.9 | | 135.6 | 145.1 | | 169.6 | 37.2 (4–10 Sept) | |
| 10 | | 96.6 | | 106.5 | 107.8 | | Down | 9.9 (4–7 Sept) | |
| 4 | 10th Aug | 138.6 | 13th Aug | 145.4 | 148.2 | 16th Aug | 170.4 | 29.0 (10–16 Aug) | SHI SH RA M |
| 5 | | 140.5 | | 151.7 | 151.7 | | 172.6 | 32.1 (10–16 Aug) | |
| 2 | | 152.8 | | 162.4 | 163.7 | | Down | 9.6 (10–13 Aug) | |
| 1 | | 128.9 | | 140.2 | 138.2 | | 160.2 | 33.3 (10–16 Aug) | |
| 8 | | 135.6 | | down | 138.4 | | 150.8 | 12.4 (13–16 Aug) | |

(continued)

**Table 4.1** (continued)

| No. of stakes | Date | Initial length of stake (in cms) | Dat e | Adjusted length of stake (in cms) | Initial length of stake (in cms) | Date | Adjusted length of stake (in cms) | Ablation rate (in cms) | GL ACI ER |
|---|---|---|---|---|---|---|---|---|---|
| 7 | | Stakes lost | | Stakes lost | Stakes lost | | Stakes lost | | |
| 3 | | 142.6 | | 152.8 | 154.2 | | Down | 10.2 (10–13 Aug) | |
| 6 | | 136.4 | | 154.0 | 149.6 | | 164.8 | 32.8 (10–16 Aug) | |
| 10 | | Stakes lost | | Stakes lost | Stakes lost | | Stakes lost | | |
| 9 | | 132.6 | | Down | 137.8 | | 146.6 | 8.8 (13–16 Aug | |

studied from 1976 to 2011. The study revealed that the glaciers have lost the area of 10.3 ± 0.03% at the rate of 0.41 ± 0.002 $km^2y^{-1}$ and niche glaciers have lost an area of 40.1 ± 0.001% at the rate of 0.04 ± 0.001 $km^2y^{-1}$. The proglacial lake of Baspa Bamak glacier has expanded from 0.020 ± 0.0004 $km^2$ in 2000 to 0.069 ± 0.001 $km^2$ in 2011 and an unnamed glacial lake has expanded from 0.09 ± 0.001 $km^2$ in 1994 to 0.10 ± 0.002 $km^2$ in 2011. Further studies, based on Landsat data (MSS, ETM + and Tm), reveal that out of total 109 glaciers in Baspa basin, 36 (32.5 ± 2.0%) are found to be covered with thick debris. The shrink in the area of glaciers has been from 227.4 ± 9.4 $km^2$ in 1976 to 186.2 ± 3.7 $km^2$ in 2011 at the rate of 1.18 ± 0.3 $km^2y^{-1}$. The retreat of 33 glaciers varied from 3.3 ± 0.03% to 3.0 ± 6.6%, i.e. 0.87 ± 0.06 km at the rate of 17.2 ± 1 $my^{-1}$ to 0.60 ± 0.04 km at the rate of 24.8 ± 0.2 $my^{-1}$, because of which debris cover on glaciers has increased by 23.5 ± 1.4 $km^2$ from 1976 to 2011 (Mir et al. 2018b). The insulating effect due to thick debris cover on the glaciers has influenced the rate of retreat of glaciers (Buri and Pellalongicciotti 2018, Shukla and Garg 2019). The change in the pattern of retreat of glaciers is primarily due to increase in temperature and decrease in precipitation. The result shows disparity in rate of retreat and deglaciation of the valley.

**Table 4.2** Snow pit measurements in the accumulation zones of Kolahoi and Shishram glaciers

| Pit | No. of samples | Length of snow sampler (in cms) | Length of sample snow column (in cms) | Weight of the snow sample column (in gms) | Density of snow sample (in gms/cm$^3$) | Water equivalent of snow sample (in gms/cm$^3$) | Cumulative water equivalent of snow pit samples (gms/cm$^2$) |
|---|---|---|---|---|---|---|---|
| *Kolahoi glacier (1984)* | | | | | | | |
| 1 | 1 | 40 | 40 | 760 | 0.43 | 17.2 | 17.2 |
| | 2 | 40 | 40 | 820 | 0.46 | 18.4 | 35.6 |
| | 3 | 40 | 40 | 990 | 0.56 | 22.4 | 58.0 |
| | 4 | 40 | 40 | 905 | 0.51 | 20.4 | 78.4 |
| | 5 | 40 | 10 | 125 | 0.52 | 2.6 | 81.0 |
| 2 | 1 | 40 | 40 | 735 | 0.42 | 16.8 | 16.8 |
| | 2 | 40 | 40 | 840 | 0.48 | 19.2 | 36.0 |
| | 3 | 40 | 40 | 925 | 0.52 | 20.8 | 56.8 |
| | 4 | 40 | 40 | 900 | 0.51 | 20.4 | 77.2 |
| | 5 | 40 | 40 | 960 | 0.55 | 22.0 | 99.2 |
| | 6 | 40 | 30 | 740 | 0.56 | 16.8 | 116.0 |
| 3 | 1 | 40 | 40 | 845 | 0.48 | 19.2 | 19.2 |
| | 2 | 40 | 40 | 872 | 0.50 | 20.0 | 39.2 |
| | 3 | 40 | 40 | 905 | 0.52 | 20.8 | 60.0 |
| | 4 | 40 | 40 | 955 | 0.54 | 21.6 | 81.6 |
| | 5 | 40 | 40 | 955 | 0.54 | 21.6 | 103.2 |
| | 6 | 40 | 40 | 980 | 0.56 | 22.4 | 125.6 |
| | 7 | 40 | 40 | 985 | 0.56 | 22.4 | 148.0 |
| | 8 | 40 | 16 | 405 | 0.58 | 9.3 | 157.3 |
| *Shishram glacier (1984)* | | | | | | | |
| 1 | 1 | 40 | 40 | 720 | 0.41 | 16.4 | 16.4 |
| | 2 | 40 | 30 | 700 | 0.46 | 13.8 | 30.2 |
| 2 | 1 | 40 | 40 | 750 | 0.43 | 17.2 | 17.2 |
| | 2 | 40 | 40 | 820 | 0.47 | 18.8 | 36.0 |
| | 3 | 40 | 40 | 860 | 0.48 | 19.2 | 55.2 |
| | 4 | 40 | 40 | 930 | 0.52 | 20.8 | 76.0 |
| | 5 | 40 | 20 | 450 | 0.51 | 10.2 | 86.2 |
| 3 | 1 | 40 | 40 | 705 | 0.42 | 16.8 | 16.8 |
| | 2 | 40 | 40 | 720 | 0.41 | 16.4 | 33.2 |
| | 3 | 40 | 40 | 960 | 0.55 | 22.0 | 55.2 |
| | 4 | 40 | 40 | 900 | 0.51 | 20.4 | 75.6 |
| | 5 | 40 | 40 | 930 | 0.53 | 20.0 | 95.6 |

(continued)

**Table 4.2** (continued)

| Pit | No. of samples | Length of snow sampler (in cms) | Length of sample snow column (in cms) | Weight of the snow sample column (in gms) | Density of snow sample (in gms/cm$^3$) | Water equivalent of snow sample (in gms/cm$^3$) | Cumulative water equivalent of snow pit samples (gms/cm$^2$) |
|---|---|---|---|---|---|---|---|
| | 6 | 40 | 40 | 940 | 0.53 | 20.0 | 115.6 |
| | 7 | 40 | 16 | 380 | 0.54 | 8.6 | 124.2 |
| 4 | 1 | 40 | 400 | 825 | 0.47 | 18.8 | 18.8 |
| | 2 | 40 | 40 | 860 | 0.48 | 19.2 | 38.0 |
| | 3 | 40 | 40 | 900 | 0.54 | 21.6 | 59.6 |
| | 4 | 40 | 40 | 940 | 0.53 | 20.0 | 79.6 |
| | 5 | 40 | 40 | 945 | 0.54 | 21.6 | 101.2 |
| | 6 | 40 | 40 | 940 | 0.53 | 20.0 | 121.2 |
| | 7 | 40 | 40 | 980 | 0.56 | 22.4 | 143.6 |
| | 8 | 40 | 40 | 985 | 0.56 | 22.4 | 166.0 |
| | 9 | 40 | 40 | 985 | 0.56 | 22.4 | 188.4 |
| | 10 | 40 | 40 | 980 | 0.58 | 23.6 | 212.0 |
| | 11 | 40 | 20 | 195 | 0.60 | 18.6 | 230.6 |

## 4.4 West Himalayan Glaciers

### *4.4.1 Drass Sub-basin Glaciers*

Drass sub-basin is the Vth order basin of IVth order Indus basin and extends from Gumri (near Zoji La pass) in the west to Kargil town in the east. It is nestled with snow-clad high mountain peaks of Himalaya in the west, Ladakh mountain range in the east and Zanskar range in the southwest. Drass town is the second coldest habited place in the world. Drass sub-basin has been studied in detail from 1965 to 2013. The area of glacier basin has changed from 187.9 km$^2$ in 1965 to 158.42 km$^2$ in 2001 and 156.65 km$^2$ in 2013, showing a decrease of 31.25 km$^2$ or 16.63% during the period from 1965 to 2013. The fragmentation increased the number of analysed glaciers from 115 in 1965 to 150 in 2013. A decrease of 29.48 km$^2$ in area, between 1965 and 2001, is due to the high degree of aridity and low precipitation, which also led to the fragmentation of glaciers and increase in the debris cover on the glaciers.

The glacier inventory, prepared on the basis of the available Survey of India topographical sheets, documents 115 glaciers confined between an altitude of 3600 and 6000 m amsl in Drass sub-basin. Out of the 115 glaciers, 81 with an area more than 0.1 km$^2$ were selected for general studies, whereas 37 glaciers out of the 81 (45.5% of the 81 glaciers), with an area more than 1 km$^2$, were selected for detail study. Out of selected 37 glaciers, 24 occupy < 3 km$^2$ area, 7 glaciers occupy the

**Table 4.3** Altitudinal extend and relative relief of the Liddar glaciers

| Glacier | Area ($km^2$) | Direction of snout from the catchment area | Length (in kms) | Breadth near snout (in m) | Extension of glacier height | | Relative relief (in m) |
|---|---|---|---|---|---|---|---|
| | | | | | Snout (m amsl) | Source area (m amsl) | |
| *West Liddar (Kolahoi group of glaciers)* | | | | | | | |
| Kolahoi | 11.9 | Northwest | 6.5 | 321 | 3690 | 5050 | 1360 |
| Musa Sabin *Qabar* | 3.0 | Southwest | 1.8 | 130 | 4000 | 4600 | 600 |
| Hoksar | 2.4 | Northwest | 1.4 | 110 | 3700 | 4400 | 700 |
| South Kolahoi | 1.4 | Southwest | 0.6 | 85 | 4000 | 4200 | 200 |
| Handut (1) | 1.5 | Northeast | 0.7 | 90 | 3900 | 4400 | 500 |
| Handut (2) | 2.4 | Northeast | 1.1 | 100 | 3950 | 4400 | 450 |
| *East Liddar (Shishram group of glaciers)* | | | | | | | |
| Shishram | 9.9 | Northwest | 4.5 | 226 | 3720 | 4950 | 1230 |
| Golal Gali | 0.7 | Northwest | 0.3 | 60 | 4100 | 4600 | 500 |
| Gratbal | 0.8 | Northwest | 0.2 | 75 | 4000 | 4600 | 600 |
| A | 0.5 | West | 0.1 | 40 | 4000 | 4600 | 600 |
| B | 0.4 | West | 0.2 | 50 | 4100 | 4600 | 500 |
| C | 0.3 | West | 0.1 | 40 | 4100 | 4600 | 500 |
| D | 0.5 | West | 0.2 | 70 | 4100 | 4600 | 500 |
| Sonsar | 6.0 | Northwest | 2.5 | 230 | 3700 | 4500 | 800 |
| Dhudhal | 3.8 | Northwest | 1.6 | 200 | 3800 | 4800 | 1000 |
| Tuliyan | 4.1 | Northwest | 1.2 | 200 | 3700 | 4600 | 900 |
| Manipal | 2.6 | Northwest | 1.1 | 95 | 4100 | 4800 | 700 |

area between 3 $km^2$ and 6 $km^2$, 3 glaciers occupy between 6 $km^2$ and 9 $km^2$ area, one glacier occupy between 9 $km^2$ and 12 $km^2$, and two glaciers occupy > 12 $km^2$ area.

The largest glacier in the basin has an area of 15.14 $km^2$. A large number of glaciers having an area > 0.1 $km^2$ clearly suggest that small-sized and niche glaciers dominate the Drass sub-basin. The mean size (0.9 $km^2$) of glaciers in the Drass sub-basin is close to the glaciers of other basins / sub-basins of Himalaya, such as Ravi (0.6 $km^2$), Shyok (1.4 $km^2$), Chenab (1.1 $km^2$), Bhagirathi (1.3 $km^2$), Ganga (1.1 $km^2$) and Brahmaputra (1.2 $km^2$). The glaciers of Drass sub-basin, more or less, are distributed equally in all geographical directions, but 66% of the glaciers with preferred orientation to north, northeast and northwest.

### 4.4.2 Long-Term Monitoring of Glaciers

Eighty-one glaciers of Drass sub-basin were monitored for long term for the period 1965–2001 and 2001–2013. The study of change in area of glaciers shows that 10 glaciers experienced gain in the area (Table 4.4), 13 glaciers lost more than 50% of the area, 18 glaciers lost from 25 to 50% of the area and the remaining glaciers had a marginal or no loss in the area. Realizing that the change in area of glaciers during observation time series of 50 years is variable, the monitoring on an annual basis was undertaken to study the change in area of glaciers. The study showed that five glaciers vacated the area at the rate of more than 10,000 $m^2y^{-1}$, 5 glaciers experienced the loss of area at the rate of 5000–10,000 $m^2y^{-1}$, 9 glaciers vacated the area at the rate of 2500–5000 $m^2y^{-1}$ and remaining lost the area at the rate of 15–1500 $m^2y^{-1}$. The melting of most of the small-sized and niche glaciers, confined to higher elevations of Drass sub-basin, is affected by the solar radiation. The annual change in area of each glacier ranges from 0.12 to 1.7% of the total area of glacier. Based on the difference between two digital elevation models applied to calculate the regional mass balance of glaciers in the Hindu Kush Himalaya, Sam et al. (2018) observed highly heterogeneous spatial pattern of changes in glaciers elevation showing thinning of ice and high ablation rate at the tongues of glaciers covered with debris.

The Space Applications Centre (SAC), Ahmedabad, recently mapped 5234 Himalayan glaciers, of which 2143 glacier bodies from high Himalaya and Karakoram, including the ones draining into Indus basin and covering an area of 18,343.39 $km^2$, were selected to assess changes in the area of glaciers for the time period from 2001 to 2017. Results reveal that in Himalayan region covering an area of 11,451.53 $km^2$ there is gain of 0.026% in the area of glaciers and loss of 1.44%

**Table 4.4** Relative change in altitudinal position of snout and gain in the area of glaciers of Drass sub- basin between 1965 and 2001

| Glaciers Id | Orientation | 2001 | | 1965 | | Gain in area |
|---|---|---|---|---|---|---|
| | | Altitude position of snout (m) | Area (in $km^2$) | Altitude position of snout (m) | Area (in $km^2$) | |
| 43 N15 03 | NW | 3990 | 3.83 | 3880 | 3.33 | 0.5 |
| 43 N15 04 | NW | 4480 | 1.96 | 4360 | 1.55 | 0.41 |
| 43 N11 13 | N | 4760 | 0.23 | 4710 | 0.20 | 0.03 |
| 43 N11 42 | NE | 4600 | 0.23 | 4590 | 0.21 | 0.02 |
| 43 N07 62 | SE | 4520 | 0.70 | 4540 | 0.68 | 0.02 |
| 43 N07 63 | SE | 4420 | 1.56 | 4400 | 1.54 | 0.02 |
| 43 N07 64 | NE | 3830 | 4.01 | 3790 | 3.68 | 0.33 |
| 43 N11 102 | NE | 4580 | 0.42 | 4600 | 0.40 | 0.02 |
| 43 N11 113 | E | 4760 | 0.19 | 4740 | 0.17 | 0.02 |
| 43 N07 140 | SW | 4740 | 0.17 | 4720 | 0.16 | 0.01 |

in the area of glaciers over a span of 17 years from 2001 to 2017. The change in glaciers varies from + 0.026% to – 0.76%, including the ones draining into Indus basin, with a variable rate of 0.76% for the glaciers of sub-basin located on northeast of northwest Himalaya flowing Indus River (Jammu & Kashmir, and Ladakh), 2.2% for the glaciers of Chenab and Sutlej basin and 0.84% for the glaciers of Ganga basin (Bahuguna et al. 2021; Patel et al. 2018).

### *4.4.3 Relative Change in Snout Position of Large-Sized Glaciers*

Eleven large-sized glaciers show considerable shift in the position of snouts / terminus during the period from 1965 to 2001. The shift in position has caused decrease in the length as well as the area of glaciers. In case of four glaciers, the vacation of area of glaciers is between 15 and 21%. Similarly, in case of another four glaciers the position of snouts shifted uphill showing maximum decrease in the length of the glaciers but a moderate decrease of < 10% in the area.

The overall vacation of area of 11 large-sized glaciers from 1965 to 2001 shows that one of the glaciers shrunk at the rate of 25,000 $m^2y^{-1}$, four glaciers vacated the area at the rate of 10,023 $m^2y^{-1}$, three glaciers vacated the area at the rate of 3920 $m^2y^{-1}$ and three at rate of 1478 $m^2y^{-1}$. Further, 13 large-sized glaciers of different dimensions got fragmented into nearly 30 glaciers during the period from 1965 to 2001, with no noticeable change in the area as well as the length of the glaciers, but with high rate of ablation of glacier tongues covered with debris. Machoi glacier, in Drass sub-basin close to Zoji La pass, shows the vacation of area since 1965 wherein it has lost 0.62 $km^2$ between 1965 and 2013, nearly 50 years, except in 2001 and 2013 when the glacier has vacated 0.09 $km^2$ of the area. The mass balance study of Machoi glacier during 2011–2014 shows a positive net balance of $+ 0.16 \pm 0.19$ $my^{-1}$.

The study of change in the area of glaciers carried out for the period from 1962 to 2001 across the Himalaya shows that surface change in the glacier area, particularly across northwest Himalaya, is a factor of spatial distribution of the glaciers. The glaciers in Parvati valley retreated by ~ 0.56%$y^{-1}$, in Zanskar by ~ 0.23%$y^{-1}$, in Gori Ganga by ~ 0.48%$y^{-1}$, in Jankar Chhu watershed (Lahaul) 7.5 ± 2.2% (0.2 ± 0.1% $y^{-1}$) and in Bhagirathi by ~ 0.35%$y^{-1}$ (Das and Sharma 2019; Das 2022). Based on the study of satellite imageries, exposed-ice area of Siachen glacier (East Karakoram) has decreased by 1.3% while the debris-covered area has increased by 10.5% from 1980 to 2014, with the debris-covered tongue remaining stable over the period (Agarwal et al. 2017).

The glaciers in Himalayan range show variable shrinkage, in general, for last 5–6 decades with rates ranging from – 0.07% $y^{-1}$ for Kimoshung glacier, Tibet region (1974–2015), to – 1.38%$y^{-1}$ for Ganju La glacier, Bhutan (2004–2014), whereas glaciers in the Karakoram range, on the contrary, showed a slight shrinkage or stable since the mid-nineteenth century (Azam et al. 2018 and references herein). Based on the comparative study of Survey of India topographical sheets (1962) and LISS II images (2001/04), a higher rate of retreat of ~ 0.8%$y^{-1}$ for Bhaga basin glaciers, ~ 0.5%$y^{-1}$ for Warwan Bhut region (Chenab basin) glaciers, ~ 0.5%$y^{-1}$ for Chandra basin glaciers and ~ 0.3%$y^{-1}$ for Samudra Tapu glacier (Chandra river basin) is noticed (Das 2022 and references herein) with a note of caution that there could be over estimation of glacier area cover in the topographical sheets. Menthosa glacier (Lahaul Himalaya) receded on an average rate of $5.9 \pm 0.4$ $my^{-1}$ between 1965 and 2016, losing 0.09 $km^2$ ice in the frontal section with continuous recessional pattern and substantial glacier degeneration over the past one decade (2006–2017) (Prakash et al. 2022). Significantly higher rate of recession of $70.9 + 3.1$ $my^{-1}$ between 1973 and 1989 led to the disintegration of Parvati glacier (NW Himalaya) (Chand et al. 2020).

### 4.4.4 Steady-State Phase in Area and Snout Position of Glaciers

The study of satellite images of 150 glaciers available for the time series from 2001 to 2013 reveals that 120 glaciers measuring between < 1 $km^2$ and > 10 $km^2$ do not show any change in (Table 4.5). The glaciers are distributed, more or less, in all geographical directions with majority (61.2%) of glaciers oriented to N, of which 21% are oriented to NW, 20.2% to N and 20% to NE. The remaining 30 glaciers show marginal change in the areas and length for the time series from 2001 to 2013. The glaciers are orientated predominantly to northwest (41%) followed by northeast and north (18.75% each). Eighteen glaciers have lost less than 0.05 $km^2$ of the area, six glaciers have lost between 0.05 and 0.1 $km^2$, two glaciers have lost between 0.1 and 0.2 $km^2$, and two glaciers vacated more than 0.2 $km^2$ of the area. The average snowline of the basin is at $5009 \pm 47$ m amsl.

## 4.5 Suru-Zanskar Valley

The Suru-Zanskar valley is a vast mountainous region occupying southern part of the Ladakh Himalaya and lying between High Himalaya range in the southwest and Indus basin in the northeast. Tectonically, it represents the northwest extension of Spiti basin. Suru-Zanskar valley is Vth order basin of IVth order Indus basin and extends between Suru-Zanskar in the west and Padum in the east. It has thirty-six sub-basins, out of which nineteen are glaciated, housing nearly 768 glaciers and covering an area of 1221.43 $km^2$. It is encircled by ridge crest of Pensi La pass and Durung Drung glacier in northwest, Umasi La pass (5342 m amsl) and Hagshu La pass in the west and southwest, Yara La pass (5697 m amsl) in the east and Baralacha La

**Table 4.5** Glaciers of Drass sub-basin showing no change in the area between 2001 and 2013

| S. No. | Glaciers Id | Orientation | Area (in $km^2$) | |
|---|---|---|---|---|
| | | | 2013 | 2001 |
| 1 | 43 N15 01 | N | 0.15 | 0.15 |
| 2 | 43 N15 03 | NW | 3.83 | 3.83 |
| 3 | 43 N15 05 | N | 1.15 | 1.15 |
| 4 | 43 N11 06 | NE | 2.44 | 2.44 |
| 5 | 43 N11 08 | W | 2.25 | 2.25 |
| 6 | 43 N12 09 | NW | 10.1 | 10.1 |
| 7 | 43 N11 10 | W | 2.79 | 2.79 |
| 8 | 43 N11 11 | W | 2.09 | 2.09 |
| 9 | 43 N11 12 | NW | 1.47 | 1.47 |
| 10 | 43 N11 13 | N | 0.23 | 0.23 |
| 11 | 43 N11 14 | NW | 0.1 | 0.1 |
| 12 | 43 N11 15 | NW | 0.08 | 0.08 |
| 13 | 43 N12 16 | NW | 1.18 | 1.18 |
| 14 | 43 N12 17 | E | 0.3 | 0.3 |
| 15 | 43 N12 18 | SW | 0.19 | 0.19 |
| 16 | 43 N12 19 | SW | 0.15 | 0.15 |
| 17 | 43 N12 20 | N | 0.26 | 0.26 |
| 18 | 43 N12 21 | N | 0.47 | 0.47 |
| 19 | 43 N12 22 | N | 0.21 | 0.21 |
| 20 | 43 N12 23 | NW | 0.09 | 0.09 |
| 21 | 43 N12 27 | NW | 0.21 | 0.21 |
| 22 | 43 N12 29 | SE | 0.72 | 0.72 |
| 23 | 43 N12 30 | SE | 0.53 | 0.53 |
| 24 | 43 N12 33 | NE | 0.21 | 0.21 |
| 25 | 43 N12 34 | NW | 0.08 | 0.08 |
| 26 | 43 N12 37 | W | 0.49 | 0.49 |
| 27 | 43 N12 38 | S | 0.45 | 0.45 |
| 28 | 43 N11 42 | NE | 0.23 | 0.23 |
| 29 | 43 N11 46 | N | 0.74 | 0.74 |
| 30 | 43 N11 47 | NW | 0.42 | 0.42 |
| 31 | 43 N11 48 | SE | 0.07 | 0.07 |
| 32 | 43 N11 49 | E | 2.28 | 2.28 |
| 33 | 43 N11 50 | NE | 0.94 | 0.94 |
| 34 | 43 N11 51 | S | 1.38 | 1.38 |
| 35 | 43 N11 52 | S | 0.09 | 0.09 |
| 36 | 43 N11 53 | S | 0.21 | 0.21 |

(continued)

**Table 4.5** (continued)

| S. No. | Glaciers Id | Orientation | Area (in $km^2$) | |
|---|---|---|---|---|
| | | | 2013 | 2001 |
| 37 | 43 N11 54 | S | 0.24 | 0.24 |
| 38 | 43 N11 55 | N | 0.1 | 0.1 |
| 39 | 43 N11 56 | NW | 0.23 | 0.23 |
| 40 | 43 N07 60 | NE | 0.23 | 0.23 |
| 41 | 43 N07 61 | NE | 0.13 | 0.13 |
| 42 | 43 N07 65 | NE | 1.41 | 1.41 |
| 43 | 43 N07 66 | SE | 2.08 | 2.08 |
| 44 | 43 N07 67 | E | 1.6 | 1.6 |
| 45 | 43 N07 68 | S | 0.1 | 0.1 |
| 46 | 43 N07 69 | S | 0.1 | 0.1 |
| 47 | 43 N07 70 | SE | 0.13 | 0.13 |
| 48 | 43 N07 71 | N | 2.01 | 2.01 |
| 49 | 43 N07 72 | NW | 2.26 | 2.26 |
| 50 | 43 N07 73 | S | 0.77 | 0.77 |
| 51 | 43 N07 74 | NW | 0.39 | 0.39 |
| 52 | 43 N07 75 | E | 1.5 | 1.5 |
| 53 | 43 N07 76 | NW | 0.53 | 0.53 |
| 54 | 43 N07 77 | N | 2.13 | 2.13 |
| 55 | 43 N07 78 | NE | 0.33 | 0.33 |
| 56 | 43 N07 79 | NE | 1.56 | 1.56 |
| 57 | 43 N07 81 | NE | 0.75 | 0.75 |
| 58 | 43 N06 82 | E | 0.41 | 0.41 |
| 59 | 43 N06 83 | NE | 0.56 | 0.56 |
| 60 | 43 N06 85 | N | 0.37 | 0.37 |
| 61 | 43 N06 86 | N | 0.25 | 0.25 |
| 62 | 43 N06 87 | N | 0.04 | 0.04 |
| 63 | 43 N06 88 | N | 0.28 | 0.28 |
| 64 | 43 N06 89 | NE | 0.2 | 0.2 |
| 65 | 43 N06 90 | NE | 0.09 | 0.09 |
| 66 | 43 N06 91 | N | 0.14 | 0.14 |
| 67 | 43 N06 92 | NE | 0.14 | 0.14 |
| 68 | 43 N10 93 | NW | 0.61 | 0.61 |
| 69 | 43 N10 94 | NW | 0.19 | 0.19 |
| 70 | 43 N10 95 | NW | 0.08 | 0.08 |
| 71 | 43 N11 96 | N | 0.04 | 0.04 |

(continued)

**Table 4.5** (continued)

| S. No. | Glaciers Id | Orientation | Area (in km$^2$) | |
|---|---|---|---|---|
| | | | 2013 | 2001 |
| 72 | 43 N11 97 | E | 0.19 | 0.19 |
| 73 | 43 N10 98 | S | 0.31 | 0.31 |
| 74 | 43 N11 99 | E | 0.26 | 0.26 |
| 75 | 43 N11 100 | NE | 0.32 | 0.32 |
| 76 | 43 N11 101 | N | 0.36 | 0.36 |
| 77 | 43 N11 102 | NE | 0.42 | 0.42 |
| 78 | 43 N11 103 | NE | 0.16 | 0.16 |
| 79 | 43 N11 104 | NE | 0.12 | 0.12 |
| 80 | 43 N10 105 | SE | 0.87 | 0.87 |
| 81 | 43 N14 106 | W | 0.09 | 0.09 |
| 82 | 43 N14 107 | N | 0.14 | 0.14 |
| 83 | 43 N14 108 | S | 0.13 | 0.13 |
| 84 | 43 N14 109 | S | 0.27 | 0.27 |
| 85 | 43 N14 110 | E | 0.27 | 0.27 |
| 86 | 43 N11 111 | NW | 0.09 | 0.09 |
| 87 | 43 N11 112 | E | 0.2 | 0.2 |
| 88 | 43 N11 113 | E | 0.19 | 0.19 |
| 89 | 43 N11 115 | N | 0.34 | 0.34 |
| 90 | 43 N11 116 | N | 0.12 | 0.12 |
| 91 | 43 N07 118 | W | 0.45 | 0.45 |
| 92 | 43 N07 120 | NE | 0.15 | 0.15 |
| 93 | 43 N07 121 | NE | 0.35 | 0.35 |
| 94 | 43 N07 122 | SE | 0.09 | 0.09 |
| 95 | 43 N07 123 | E | 0.15 | 0.15 |
| 96 | 43 N07 124 | SE | 1.5 | 1.5 |
| 97 | 43 N07 125 | SW | 0.63 | 0.63 |
| 98 | 43 N07 126 | SE | 0.23 | 0.23 |
| 99 | 43 N07 127 | SW | 0.36 | 0.36 |
| 100 | 43 N07 128 | SW | 0.58 | 0.58 |
| 101 | 43 N07 129 | S | 0.46 | 0.46 |
| 102 | 43 N07 130 | NW | 0.23 | 0.23 |
| 103 | 43 N07 131 | NW | 3.78 | 3.78 |
| 104 | 43 N07 132 | NE | 2.57 | 2.57 |
| 105 | 43 N07 133 | NW | 0.02 | 0.02 |
| 106 | 43 N07 134 | NW | 0.36 | 0.36 |

(continued)

**Table 4.5** (continued)

| S. No. | Glaciers Id | Orientation | Area (in km$^2$) | |
|---|---|---|---|---|
| | | | 2013 | 2001 |
| 107 | 43 N11 135 | N | 0.14 | 0.14 |
| 108 | 43 N11 136 | N | 0.92 | 0.92 |
| 109 | 43 N07 137 | SE | 0.16 | 0.16 |
| 110 | 43 N07 138 | E | 0.2 | 0.2 |
| 111 | 43 N07 140 | SW | 0.17 | 0.17 |
| 112 | 43 N07 141 | N | 1.43 | 1.43 |
| 113 | 43 N07 142 | N | 1.01 | 1.01 |
| 114 | 43 N07 143 | NW | 0.35 | 0.35 |
| 115 | 43 N07 144 | NW | 0.15 | 0.15 |
| 116 | 43 N07 145 | S | 0.19 | 0.19 |
| 117 | 43 N11 146 | NE | 0.13 | 0.13 |
| 118 | 43 N11 147 | N | 0.1 | 0.1 |
| 119 | 43 N11 148 | NE | 0.38 | 0.38 |
| 120 | 43 N11 149 | NE | 0.17 | 0.17 |

pass (4650 m amsl) in the southeast. The Zanskar mountain houses Haptal, Yaranchu, Mulung, Kanthang, Denyai, Shimling, Chabber, Gompi, Pholoklo, Khapang, Tamasi and Churlup glaciers facing northeast and corresponds to the general alignment of the mountain rim. Small-sized glaciers of 1–2 kms in size are found to the north of the valley and are oriented to south and southwest direction.

Zanskar mountain, a denuded mountain, consists of sedimentary rocks of grey and green siltstone with occasionally slate. The rocks preserve sedimentary structures within the purple-coloured fine-grained sandstone and covers an area of 1568.43 km$^2$ which is 22.40% of the total area. The rocks exposed at higher elevation of the uplands are racked up into folds giving rise to protracted broad valleys having depression along the slopes that house large number of Zanskar glaciers in south and southwest. The glaciers have more accumulation in winter and receive less insulation in summer. The Zanskar valley is filled with a large system of glaciers that extend deep into Stod-Lungnuk and Zanskar gorge to the southeast of Zanskar valley. The region is drained by Zanskar River having two main tributaries, namely Doda and Lungnak (confluence of Kurgiakh and Tsrap Chu) rivers. Suru River, commencing from Pensi La pass, traverses along Suru valley from west to east to join Indus and Zanskar rivers near village Nimmu near Leh and then turns towards Skardu.

### 4.5.1 Long-Term Monitoring of Glaciers

The inventory of glaciers prepared based on the available Survey of India topographic sheets, document 268 glaciers in Zanskar valley out of which 131 glaciers are housed in High Himalaya and 137 in Zanskar mountains. Majority (98 in number) of the glaciers are > 1 km in length and 90 glaciers are 1 km to 2 kms in length. Fifty-eight glaciers (60%) are in the range of > 5 km$^2$, 19 glaciers (19.8%) are between 5 km$^2$ and 15 km$^2$, 8 (8.3%) glaciers are between 15 km$^2$ and 30 km$^2$, 4 (4.2%) glaciers are between 30 km$^2$ and 45 km$^2$, 2 (2.1%) glaciers are between 45 km$^2$ and 60 km$^2$ and 2 (2.1%) glaciers are greater than 60 km$^2$ in the area, including Durung Drung glacier covering an area of 72.14 km$^2$. The glaciers in Suru-Zanskar valley are distributed in all geographical directions with 22% of glaciers with orientation to north, 21% oriented to northeast and 19% oriented to northwest.

The outlines of glaciers drawn on the Survey of India topographic sheets of 1965 of the scale 1:50,000 were overlain on multi-satellite images of Landsat and IRS LISS III satellite data to detect the change in glaciers applying standard ratio technique for on-screen delineation, and comparison of DEM of different dates for the period from 2001to 2013. Comparative studies were restricted to 96 glaciers for the years 1965, 2001 and 2013 (Table 4.6) after making overlay adjustments. The studies suggest degeneration of Suru-Zanskar glaciers during the period between 1969 and 2001, leading to change in temporal extent of the glaciers from 855.27 km$^2$ in 1965 to 783.06 km$^2$ in 2001 and further to 777.29 km$^2$ in 2013. There is a loss of mass of 72.21 km$^2$ during 36 years from 1965 to 2001. Compared to earlier reports, glaciers show a low rate of retreat and loss of less mass for the time scale from 2001 to 2013 due to warmer late winter season and rainy summers. The fragmentation of the large-sized glaciers to small-sized ones took place in early period (1965–2001) and exposed fresh fragmented glaciers on the southwest slopes, with supply of debris from potential slopes leading to change in the extent of debris cover. The result of recession and loss of area of Suru-Zanskar glaciers are in agreement with the studies on 252 glaciers of Suru basin applying GIS environment and supplemented with field work by Wadia Institute of Himalayan Geology, Dehradun. The studies report shrinking of glaciers by ~ 6 ± 0.02% at an average rate of retreat of 4.3 ± 1.02 my$^{-1}$ because of local warm and wetter climate conditions, with an increase in debris by 62% from 1971 to 2017 (Shukla et al. 2019). Ali et al. (2017) studied the temporal changes along the glacier valleys of Pensilungpa, Durung Drung, Kangi, Hongshu and one unnamed glacier of Zanskar basin with the help of multi-temporal satellite datasets for the time period of 1980–2000 and 2000–2011 and reported that mean ELA of glaciers increased, and area of glaciers reduced from 291.98 to 250.96 km$^2$.

**Table 4.6** Long-term changes in area (in km$^2$) of selected glaciers of Suru-Zanskar valley

| Glaciers | Area (in km$^2$) calculated | | | Change y$^{-1}$ in area (in km$^2$) |
|---|---|---|---|---|
| | SOI sheets of 1965 | LISS III image of 2001 | LISS III image of 2013 | |
| Durung Drung | 72.53 | 66.28 | 67.0 | 0.104 |
| Haskira | 18.0 | 16.0 | 12.0 | 0.125 |
| Kange | 35.9 | 33.5 | 33.0 | 0.042 |
| Hagshu | 56.0 | 54.0 | 53.4 | 0.054 |
| Sumche | 16.0 | 14.0 | 13.0 | 0.063 |
| Lochan | 8.5 | 7.6 | 7.3 | 0.025 |
| Mulung | 24.5 | 23.9 | 23.5 | 0.021 |
| Kanthang | 21.6 | 18.4 | 18.0 | 0.075 |
| Haptal | 34.0 | 31.0 | 30.3 | 0.077 |
| Yuranchu | 27.0 | 24.6 | 24.0 | 0.063 |
| Shimiling | 6.5 | 5.5 | 5.5 | 0.021 |

### 4.5.2 *Short-Term Monitoring and Current Behaviour of Glaciers*

Analyses of satellite imageries of Suru-Zanskar glaciers have revealed that 105 glaciers of different dimensions do not show any change in the area between the period 2001 and 2013. Out of 105 glaciers, 71 glaciers are less than 5 km$^2$, 18 glaciers are from 5 to 10 km$^2$, 8 glaciers are from 10 km$^2$ to 20 km$^2$, 3 glaciers are from 20 to 30 km$^2$, 4 glaciers are from 30 km$^2$ to 45 km$^2$ and one glacier is from 45 to 60 km$^2$ in area. The glaciers are heterogeneously oriented in different directions but most (60.2%) of them are oriented to the north. However, recent studies carried on to investigate the health of Dalung (Suru), Padum (Zanskar) and Pensilungpa (Zanskar) glaciers during the short period from 2015 to 2019 indicate secular retreat of glaciers at the rate of 4–6 my$^{-1}$ as a result of recent increase in temperature (Mir and Majeed 2018). The Dalung and Padum glaciers, showing a marginal variation, have shrunk by $2.6 \pm 0.56$ km$^2$ at the rate of $0.05 \pm 0.01$ km$^2$y$^{-1}$ and $3.4 \pm 0.65$ km$^2$ at the rate of $0.06 \pm 0.01$ km$^2$y$^{-1}$, respectively.

## 4.6 Mass Balance Studies

Mass balance is the most meaningful description of the hydrological cycle of glaciers, and the best means of estimating the movement of glacier and understanding the glacial processes. The mass balance studies have been extensively applied to study the movement of glaciers and record fluctuation in the glaciers of Alps in Europe and Rockies in America. Mass balance is the theoretical state of a glacier in which accumulation and loss of snow/ice by ablation are measured using various methods to assess the status of health of a glacier. The equilibrium line altitude (ELA) of a glacier is defined where the vertical mass balance profile (net mass change) is zero. The equilibrium line separates the ablation from accumulation zone of the glacier surface. The annual glacier mass balance study provides most accurate understanding of the fluctuation in glaciers in response to the prevailing climate conditions. Reconstructing long-term annual glacier mass balance with respect to the time series coupled with meteorological variable data provides most comprehensive insight to the climate change.

The field measurements of mass balance of Kolahoi-Shishram and Naradu glaciers in northwest Himalaya, and Machoi and Durung Drung glaciers in west Himalaya were undertaken to study the rate of net accumulation and ablation. The net accumulation or winter balance was measured, in terms of water equivalent volume (w.e.v.), in the months of April/May of 1983–1984 for the Kolahoi-Shishram and 2000–2001, 2001–2002 and 2002–2003 for the Naradu glaciers. The net ablation or summer balance was measured, in terms of w.e.v., using ablation stakes fixed from snout to equilibrium line near accumulation zone, during summer months from May to September of 1983 and 1984 in case of Kolahoi-Shishram glaciers and during summer months of 2000, 2001, 2002 and 2003 in case of Naradu glacier (Fig. 2.1a–c). The value was subtracted from the net accumulation of the Kolahoi-Shishram and the Naradu glaciers to compute the mass balance. The snow depth was measured by sounding through fresh snow to the glacier ice surface using 4-m-long metallic rod with pointed end. The pits in snow were excavated at some of the sounding sites on glacier body with shovels to confirm the total accumulation of snow, density of the snow and to measure the water equivalent of snow/ice in the field. The annual measurement of net accumulation of snow, by opening pits in accumulated residual snow, was carried out at the end of ablation season, i.e. September, at the equilibrium line zones at 4400–4500 m amsl of Kolahoi-Shishram glacier and at 4880–4910 m amsl of Naradu glacier. Net accumulation, at both the glaciers, varied between 50 and 100 cm in water equivalent. The surface ablation of the glaciers, measured with the help of ablation stakes fixed in the months of September/October, gave a value of 3.1–6.4 cm. The net balance of Kolahoi glacier is recorded as $-\ 7.95 \times 10\ m^3$ and the rate of retreat as 4.5 $my^{-1}$, net balance of Shishram glacier is recorded as $-\ 7.04 \times 10\ m^3$ and the rate of retreat as 3.5 $my^{-1}$ with reference to the snout position of 1983. The Naradu glacier showed variable rate of retreat between 4.3 and 3.9 m with specific balance of − 0.28, − 0.35 and − 0.40 for the years 2000–2001, 2001–2002 and 2002–2003, respectively. As a consequence of the fluctuation in net negative balance of the glacier, there is retreat of snout and vacation of glacial area to 663, 307 and 586 $m^3$ during time series 2000–2001, 2001–2002 and 2002–2003, respectively.

Accumulation stakes of 6 m length were fixed in the month of September/October with the help of Heucke ice drill in the accumulation zone of Durung Drung glacier to measure the residual snow cover. The ablation stakes were fixed to measure the ablation at different elevations from snout to the equilibrium line of the glacier. The measurement of thickness of residual winter snow in water equivalent of Durung

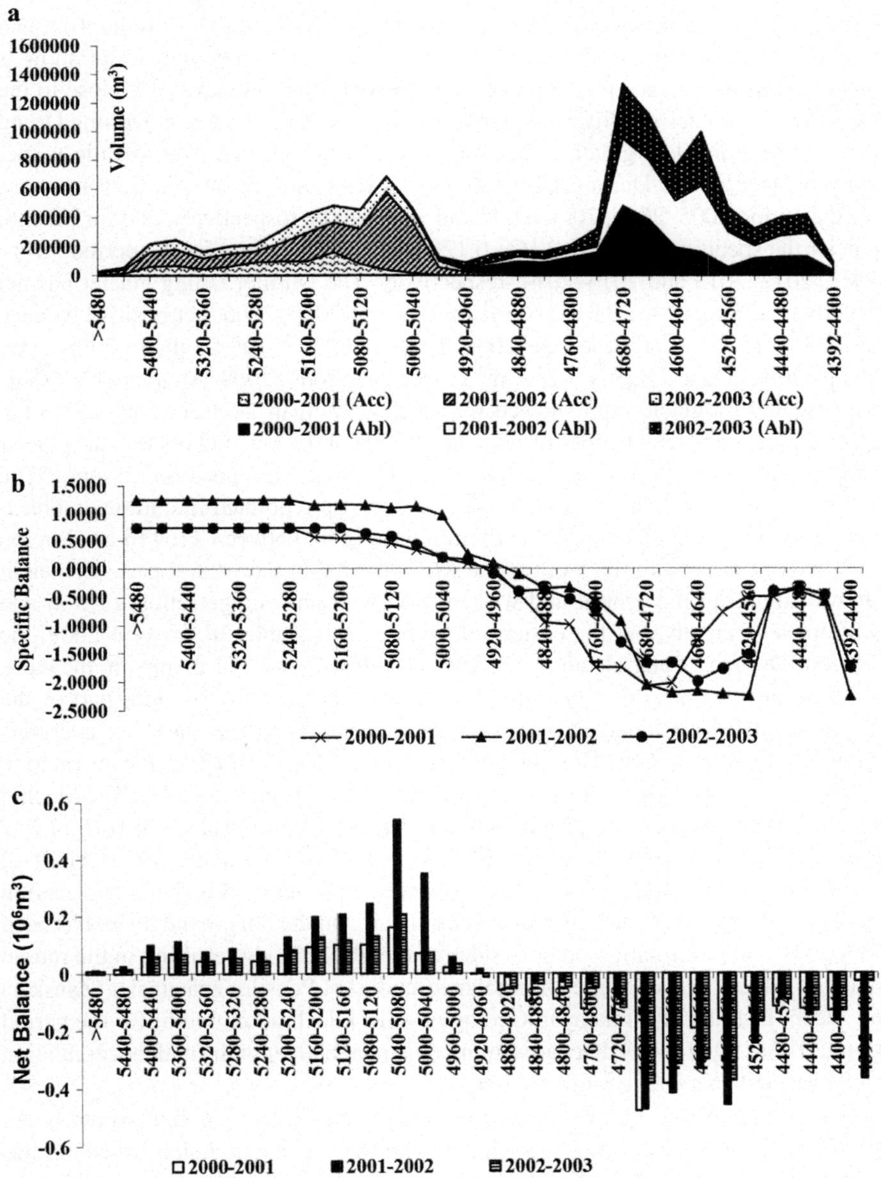


**Fig. 4.1** Graphic presentation of **a** elevation vs accumulation and ablation volume **b** specific balance and **c** net balance of Naradu glacier from 2000 to 2003

Drung glacier is done during the field season from 2006 to 2008 (Table 4.7), and in case of Machoi glacier from 2012 to 2014 (Table 4.7). The density of snow is measured from the snowpack dug out from pits at different elevations close to the ELA. The snowpack density ranged from 0.39 gm/cc to 0.49gm/cc at Durung Drung glacier and from 0.36 gm/cc to 0.45gm/cc at Machoi glacier. The specific winter mass balance of the Durung Drung glacier is 0.48 m, 0.59 m and 0.56 m w.e.v. for the period 2005–2006, 2006–2007 and 2007–2008, respectively, and for Machoi glacier the specific balance is 0.166, 0.120 and 0.174 m w.e.v. for the period 2011–2012, 2012–2013 and 2013–2014, respectively. The Durung Drung glacier has net negative balance of $-7.34 \times 10^3$ $m^3$ for the year 2005–2006, net positive balance of $2.83 \times 10^2 \times m^3$ for the year 2006–2007, $-6.77 \times 10^2 \times m^3$ in 2008–2009 and positive balance $1.48 \times 10^2 \times m^3$ for the year 2007–2008. An area of 1300 $m^2$ was vacated by glacier and showed retreat of 0.77 m in 2006, in 2007–2008 the glacier gained an area of 1550 $m^2$ with an advance of 0.53 m, and in 2009 the glacier vacated an area of 960 $m^2$ with a retreat of 0.56 m in the position of snout. The present snout of the glacier is highly irregular, sharply pointed in outline and bird-foot shaped in form and situated at elevations ranging between 4164 m amsl in the east to 4169 m amsl in the west and at 4176 m amsl in the central part. The snout shows calving in the central and western part with an average retreat of 3 m and 1.85 m, respectively, but the eastern part of snout is stable. In the year 2009, the snout of Durung Drung glacier ascended to an altitude of 4171 m amsl in the west, 4178 m amsl in the central and 4162 m amsl in the east. The eastern part of the glacier snout has remained stagnant whereas the central and west part has undergone a minor retreat of 0.76 m. The change in snout position is primarily a manifestation of variation in basal topography of the terminus region of glacier valley. The Machoi glacier has a positive net balance with cumulative specific balance of 0.16 m.w.e/year. This has resulted in the shift of ELA from 4540 m amsl in the year 2011–2012 to 4509 m amsl in 2013–2014 causing advance of glacier by 4 m from an elevation of 3656 m amsl to 3652 m amsl in the central part. On the other hand, there has been squeezing and deformation on both sides of the glacier body resulting in the retreat of 0.76–3.43 m. However, the mass balance studies of Pensilungpa glacier (Zanskar) by Wadia Institute of Himalayan Geology (WIHG, Dehradun, India) for the period from 2015 to 2019 showed negative trend in annual budget with small accumulation area ratio and an average rate of retreat of 6.7 $my^{-1}$.

The important source of moisture to nourish the glaciers in Karakoram is the Western Disturbances during the winters whereas the South Asian Monsoon has minimum or no access to the Karakoram ranges. On the other hand, western Himalaya receives precipitations from both, Western Disturbances and South Asian Monsoon, with a variation in the precipitation from south to north due to strong orographic differences and decrease in intensity from east to west (Azam et al. 2018). The studies reveal that the main portion of Tibet Plateau is experiencing rise in the surface temperature since mid-1950s (Ding et al. 2018). The thermal and mechanical forcing of Tibetan Plateau, with an average elevation of 4000 m amsl and an area of $2.5 \times 10^6$ $km^2$, impacts the regional and global climate. Wang et al. (2017) documented increase in annual temperature at 0.42 °C $decade^{-1}$ and winter increase of 0.48 °C

**Table 4.7** Rate of fluctuation (retreat/advance) in the snout position of Durung Drung and Machoi glaciers

Durung Drung glacier

| Year | Snout Position (m amsl) | | | Lateral retreat/advance (in m) | | | Area vacated/ advance (in $m^2$) | Retreat/Advance (in m) |
|---|---|---|---|---|---|---|---|---|
| | East | Centre | West | East | Centre | West | | |
| 2005 | 4164 | 4176 | 4169 | – | – | – | – | – |
| 2006 | 4164 | 4177 | 4170 | 1200 | –1450 | –1059 | –1300 (vacated) | 0.77 (retreat) |
| 2007–08 | 4162 | 4177.5 | 4170.5 | – | –750 | –450 | 1550 (advance) | 0.91 (advance) |
| 2009 | 4162 | 4178 | 4171 | – | –480 | –480 | –960 (vacated) | 0.56 (retreat) |

Machoi glacier

| Year | Snout position (m amsl) | | | Lateral retreat/advance (m) | | | Area vacated/advance ($m^2$) | Retreat/advance (in m) |
|---|---|---|---|---|---|---|---|---|
| | East | Centre | West | East | Centre | West | | |
| 2012 | 3664 | 3660 | 3670 | – | – | – | – | – |
| 2013 | 3666 | 3656.63 | 3672 | –10.0 | 3.37 | –20.0 | –200 m (lateral-vacation) | 3.37advance |
| 2014 | 3668 | 3652 | 3675 | –8.0 | 4.6 | –15.0 | –320 m (lateral-vacation) | 4.6 advance |

decade$^{-1}$ between 1979 and 2012 on Tibetan Plateau. Studies in Nepal (Karki et al. 2020) suggest winter cooling in southern lowlands and widespread warming, which is higher for maximum temperature (~0.04 °C y$^{-1}$) than for minimum temperature (~0.02 °C y$^{-1}$). The increasing trend of retreat rate and area loss in Himalayan glaciers from 1975 to 1989/1992 and 1989/1992 to 2001 is indicative of climate warming, while cooling conditions are noticed in post-2001 period (Pandey et al. 2011 and references herein).

The instant and direct response of glaciers to climate change is recorded in variations to glacier thickness, mass balance and Equilibrium Line Altitude (ELA). The ELA of Nun Kun glacier groups (Suru valley, Zanskar) varied from 4462 to 5487 (mean ELA 4979 $\pm$ 187 m amsl) to 4637–5550 m amsl (mean ELA 5062 $\pm$ 231 m amsl) from 2000 to 2020, respectively, showing an upward shift of 83 $\pm$ 22 m (4.2 $\pm$ 0.9 ma$^{-1}$) during the period (Romshoo et al. 2024b). Study of glacier changes in Sindh basin (Ahmad et al. 2022) shows increase in the mean elevation of glaciers from 4330 m amsl in 1990 to 4390 m amsl in 2018, while mean Equilibrium Line Altitude (ELA) of glaciers (> 0.5 km$^2$) shifting from 4336 m amsl to 4432 m amsl with an annual upward shift of 3.4 m y$^{-1}$. The glaciers in Upper Indus Basin of northwest Himalaya with mean elevation > 6000 m amsl show higher loss of ice thickness compared to the ones at altitudes between 5000 m amsl and 6000 m amsl due to the fact that nearly ~ 35% of the total glacier area above 6000 m amsl is oriented south compared to ~ 23% glaciers between 5000 m amsl and 6000 m amsl elevation (Abdullah et al. 2020).

Estimated ELAs of glaciers in Chitral (Hindu Kush range), Hunza (Karakoram range) and Astore (Himalayan range) basins for 2000–2018 showed upward shift from 4780 m amsl to 4980, 4580 m amsl to 4920, 4920 m amsl to 4980 m amsl and 5000 m amsl to 5020 m amsl in case of Atral, Maski-o-Shayaz, Noroghikun and Lower Tirich glaciers in Chitral basin, respectively, but lowering of ELA from 5000 m amsl to 4960 m amsl in case of Kotgaz glacier (Hayat et al. 2023). Further, Hispar, Passu, Yazgil and Mulungutti glaciers in Hunza basin showed upward shift from 4500 to 4860 m amsl and glaciers of Astore basin showed upward shift from 4420 to 5270 m amsl. The reconstruction of ELA of Qiyi glacier (Qilian mountains, Tibetan Plateau) shows upward movement from 4600 m amsl in 1970s to 4670 m amsl in 1980s that shifted further to 4970 m amsl by 2002/2003 with increase in air temperature and lastly a dramatic retreat of ELA to 5131 m amsl (Shi et al. 2023). The Asian Water Tower comprising Tibetan Plateau, Hindu Kush, Karakoram and Himalaya region has been warming since the beginning of mid-1950s showing increase by ~ 0.44 °C decade$^{-1}$ during 1979–2020 (Yao et al. 2022). Winter warming (0.46 °C decade$^{-1}$) started in 1955 and is almost double the summer warming (0.26 °C decade$^{-1}$) that started in 1986 showing pronounced seasonal and spatial differences (Yao et al. 2019).

## 4.7 Shyok-Nubra Valley Glaciers

The Shyok-Nubra valley is a tri-armed valley transected by the Nubra and Shyok rivers. Siachen glacier is the source for Nubra River. The Tibetan name of Nubra is Dumra which means 'valley of flowers'. It lies between mighty Karakoram and Ladakh mountains. Shyok-Nubra valley extends from the Tibetan Plateau called Quing Zang in China and stretches to a part of Ladakh. Karakoram terrain and Shyok-Nubra valley comprise of sedimentary, metamorphic and magmatic rocks and is tectonically very active. The valley houses about 600 glaciers (Space Applications Centre 2016) covering an area of 2178.84 $km^2$ of which 1263 $km^2$ is confined to the accumulation area and 915.84 $km^2$ to the ablation area. The per cent accumulation area of glaciers is highest in the Indus basin as compared to Ganga and Brahmaputra basins, indicating that the glaciers of Indus basin have large catchment in high Himalaya and Karakorum facilitating the stability of glaciers more than the glaciers of other basins. Indus basin has high mean accumulation and low mean ablation area (Space Applications Centre 2016), due to the predominance of Westerlies over the Indian Summer Monsoon. One hundred and fourteen glaciers of different dimensions ranging in size from less than 5 km to more than 50 km (including Siachen, the second largest glacier) were selected and studied by the authors in the first phase of investigation. Almost 52.6% of the studied glaciers are of the size less than 5 km and 31.5% of the total glaciers are between the size of 5 and 10 km. The study of 114 glaciers of Shyok-Nubra valley sub-basin suggests that 39 glaciers (34%) have gained the area, 42 glaciers (37%) have vacated the area and 33 glaciers (29%) do not show any change in the area (Table 4.8) for the years 1969, 1989 and 2001, pointing that 63% of glaciers are either stable or have gained the area. Eighty-four glaciers, out of the 114 glaciers, have been studied on a short-term basis for the time between 1989 and 2001, whereas 30 glaciers have been studied on a long-term basis for the period between 1969 and 2001.The Survey of India topographic sheets of 1969 and satellite imageries of time series between 1989 and 2001 forms the basis of the study of glaciers.

The heterogeneous glacier surfaces located in the transitional western Himalaya particularly in Chenab, Beas and Sutlej basins reveal high rate of retreat of front (>20 $my^{-1}$) of glaciers than those located in the high elevations because of the differential heating of air mass (Brahmbhatt et al. 2017). The average rate of retreat of Gangotri glacier is estimated to $17.9 \pm 0.5$ $my^{-1}$ from 1965 to 2015 with lower rate of retreat between 1965 and 1968 and increased rate during 1968–1973. The rate of retreat in recent years (2006–2015), however, was lower ($\sim 9.0 \pm 3.5$ $my^{-1}$) to the rate ($\sim 19.7 \pm 0.6$ $my^{-1}$) during 1965–2006 adding to the thickness of Gangotri and Raktavan (tributary) glacier during recent years (Bhattacharya et al. 2016). Chorabari glacier (Alaknanda catchment) receded at an average rate of 8.62 $my^{-1}$ during 1976–1990 with variable rates at different time intervals: 7.00 $my^{-1}$ during 1990–2001, 5.02 $my^{-1}$ during 2001–2012 and 10.66 $my^{-1}$ from 2012 to 2017 (Kumar et al. 2021b).

**Table 4.8** Glaciers of Shyok-Nubra valley showing no change in the area between 1989 and 2001

| S. No. | Glacier Id No | Orientation | Area (in $km^2$) | |
|---|---|---|---|---|
| | | | 1989 | 2001 |
| 1 | 52E4 012 | S | 1.703 | 1.703 |
| 2 | 52E11 004 | E | 13.400 | 13.400 |
| 3 | 52E11 005 | E | 10.736 | 10.736 |
| 4 | 52E11 006 | N | 1.443 | 1.443 |
| 5 | 52E12 005 | SSW | 1.306 | 1.306 |
| 6 | 52E12 014 | W | 4.216 | 4.216 |
| 7 | 52E12 016 | W | 1.981 | 1.981 |
| 8 | 52E16 006 | N | 0.443 | 0.443 |
| 9 | 52E8 010 | NE | 13.488 | 13.488 |
| 10 | 52E8 013 | W | 4.542 | 4.542 |
| 11 | 52F10 001 | SE | 42.698 | 42.698 |
| 12 | 52F10 005 | W | 3.679 | 3.679 |
| 13 | 52F10 006 | W | 4.493 | 4.493 |
| 14 | 52F10 009 | W | 14.172 | 14.172 |
| 15 | 52F13 002 | NW | 9.241 | 9.241 |
| 16 | 52F5 001 | NE | 3.463 | 3.463 |
| 17 | 52F5 002 | NE | 17.624 | 17.624 |
| 18 | 52F5 006 | N | 4.267 | 4.267 |
| 19 | 52F5 008 | NW | 4.617 | 4.617 |
| 20 | 52F5 009 | S | 4.335 | 4.335 |
| 21 | 52F5 014 | SW | 1.058 | 1.058 |
| 22 | 52F9 003 | W | 7.186 | 7.186 |
| 23 | 52F9 007 | SE | 5.563 | 5.563 |
| 24 | 52F9 008 | E | 4.298 | 4.298 |
| 25 | 52F9 009 | N | 1.174 | 1.174 |
| 26 | 52F9 010 | NW | 4.404 | 4.404 |
| 27 | 52F9 012 | SE | 1.478 | 1.478 |
| 28 | 52F9 013 | SSW | 48.151 | 48.151 |
| 29 | 52F9 015 | W | 3.416 | 3.416 |
| 30 | 52F9 016 | N | 1.722 | 1.722 |
| 31 | 52F9 017 | W | 1.308 | 1.308 |

### 4.7.1 Short-Term Monitoring of Glaciers

The eighty-four glaciers selected for the study on short-term basis for a period between 1989 and 2001 are dominantly (24%) oriented to the east, 16% each to the west and northwest, 12% are oriented to southeast and 8% each to north-northwest and southwest directions. Twenty-six glaciers of 84 glaciers in Nubra valley show vacation of area from 1989 to 2001. Thirty-three glaciers do not show any change in the area between 1989 and 2001, whereas 25 glaciers have gained the area from 1989 to 2001. The glaciers of Shyok-Nubra valley have either been advancing or remained stable between the time 2000 and 2008. The gradual acceleration in flow of ice of non-surging Baltoro glacier during the 2000s is another sign of a stable or growing glacier. Comparative study of satellite imageries of the period between 1969 and 1989 with that of 2001 shows considerable decline in Siachen glacier area from 994.99 $km^2$ in 1969 to 932.90 $km^2$ in 1989 to 930.37 $km^2$in 2001. Accordingly, length of the largest Siachen glacier reduced from 68.30 km in 1969 to 67.56 km in 1989 and to 67.5 km in 2001. The stable or advancing state of Karakoram glaciers is considered to be a result of increased snowfall and cooling at high altitudes (Lhakpa et al. 2022). The gain or loss in the area of glaciers in Karakoram and adjoining sub-basins is governed by the factors of latitudinal variation and local weather conditions (Bahuguna et al. 2021).

### 4.7.2 Long-Term Monitoring of Glaciers

The comparative evaluation of thirty glaciers shows that three glaciers have gained the area since 1969, eight have gained the area from 1969 to 2001 and subsequently vacated the area from 1989 to 2001, and five glaciers show vacation of the area from 1969 to 1989 followed by gain from 1989 to 2001 (Table 4.9). At the same time, seven glaciers show continuous vacation of the area from 1969 to 2001, three show gain in the area from 1969 onwards (Table 4.10), and four glaciers show vacation of the area from 1969 to 1989 and subsequently no change in the area from 1989 to 2001. Seventeen glaciers have lost the area from 2150 $km^2$ in 1969 to 2026 $km^2$ in 2001.

The detail study, to measure the terminus of glaciers precisely based on geodetic mass budget and field investigations, of the Bara Shingri, Hamtah, Ratnab, Panchi Nala glaciers of Chandra basin, Samudri Tapu glacier of Garhwal Himalaya, Parvati valley glaciers (Beas) and other 20 selected glaciers for the time from 1863 to 2016 covering a span of 150 years, revealed retreat between 1055.6 $\pm$ 21.4 m and 2898 $\pm$ 50 m with a mean rate of retreat of 28.5 $my^{-1}$ (Bhattacharya et al. 2016; Chand et al. 2017; Shukla et al. 2017; Mukerji et al. 2018; Sahu and Gupta 2020).

**Table 4.9** Glaciers of Shyok-Nubra valley showing vacation between 1969 and 1989 followed by gain in the area between 1989 and 2001

| Glacier Id No | Orientation | Area (in km$^2$) | | |
|---|---|---|---|---|
| | | 1969 (SOI) | 1989 (LiSS) | 2001(LiSS) |
| 52E3 002 | SW | 8.476 | 5.220 | 5.404 |
| 52E4 001 | NE | 9.529 | 8.824 | 8.971 |
| 52E4 002 | NE | 30.570 | 28.302 | 28.323 |
| 52E4 005 | E | 5.378 | 4.341 | 4.410 |
| 52E4 013 | SW | 0.698 | 0.489 | 0.592 |

**Table 4.10** Glaciers of Shyok-Nubra valley showing continuous gain in the area between 1969 and 2001

| Glacier Id No | Orientation | Area (in km$^2$) | | |
|---|---|---|---|---|
| | | 1969(SOI) | 1989(Landsat) | 2001(LiSS) |
| 52E3 006 | SW | 1.313 | 1.344 | 1.386 |
| 52E4 003 | NE | 35.431 | 45.221 | 45.309 |
| 52E8 007 | NW | 3.236 | 3.338 | 3.368 |

## 4.8 Summary

In the recent decade, extensive research has been carried out with focus on volumetric changes of glaciers to elucidate the current behaviour of glaciers in northwest and west Himalaya. An updated estimation of changes in glaciers are worked out using IRS LISS III images for the years 1989, 2001 and 2013 in a GIS environment to map the margins of glaciers, estimate change in the area of glaciers and changes in the thickness of glaciers at Liddar and Baspa valleys (northwest Himalaya) and Drass, Suru-Zanskar and Nubra valleys (west Himalaya). Extensive ground assessment has been ensured to match the interpretations drawn from the images and accuracy of data is attained through the application of morphometric procedures. The study, so far, shows that northwest and western Himalayan glacier basins contain large number of small-sized glaciers that recede at faster rate than the large-sized glaciers because of their location at higher altitude and latitude. The north facing glaciers show less recession compared to the south facing ones. The glaciers of Liddar and Baspa valleys in northwest Himalaya receded at faster rate of 10–30 my$^{-1}$ compared to the glaciers of west Himalaya that shows a phase of stability with a marginal loss (+ 0.02 to – 0.57%) in area of glaciers during the time from 2001 to 2013 compared to the time from 1965 to 2000. The departure between the rate of recession of northwest and west Himalaya is a factor of increase in temperature during late winter season leading to sublimation, and wet summer season which, otherwise, used to be dry due to acute aridity. The benchmark glaciers, e.g. Machoi, Kangriz, Durung Drung and Siachen, experienced reduction in surface areas leading to thinning and deformation near the snout, and the snout showed periodical advance and marginal retreat to attain stability.

# Chapter 5
# Glacial Landforms of Northwest Himalaya—Process and Origin

**Abstract** Based on the thermal profile of the glacier, the glaciers of Sikkim, Uttarakhand, Himachal Pradesh and Jammu and Kashmir are classified to warm type of glaciers whereas glaciers of Ladakh and adjoining region are classified to cold type of glaciers. The glaciers in Karakoram Himalaya, classified as surge glaciers, are characterized by rapid advance of snout due to the transfer of mass to ablation zone in short period of time. The cycles of Pleistocene glacial and interglacial periods have been responsible in facilitating the glacial erosion during the glacier advance by the processes of internal deformation, basal sliding and bed deformation in the body of glacier while transport and deposition of paraglacial and/or nonglacial sediments have been predominant during the glacier retreat. The access of rainwater, if any, to the bed of glacier also promotes the basal sliding, glacial erosion and evacuation of sediments. The transportation of sediments at the bottom and margins of glacier is subject to relationship between glacier cover, sliding velocity, glacier thermal regime, glacier hydrology, subglacial and proglacial sediment cascade and erosion rates. The 'hydraulic jacking' of overlying ice reduces the contact between sole of glacier and the rock bed causing complexities in the subglacial dynamics of glacier. The study of erosional and depositional landforms of Kolahoi, Zanskar, Suru, Satopanth, Chorabari, Pimu, Menthosa, Dhauliganga, Bhagirath Kharak glacier valleys have been useful in building the Pleistocene landscape.

**Keywords** Surge glacier · Basal sliding · Bed deformation · Spring events · Hydraulic jacking · Glacier thermal regime · Fluted moraines · Ice-cored moraines · Roche moutonnée

In the geological past, earth's surface was covered several times by snow/ice of glaciers. The recent glaciation period beginning at ~ 2 Ma (Pleistocene period) ended at ~ 11 kyr (beginning of Holocene period). The change in climate since the origin of the planet earth has been the prime cause of glaciation. The interlink between climate fluctuations and glaciation is because:



R. K. Ganjoo and M. N. Koul, *The Himalayan Glaciers, Climate Change and Society*,
https://doi.org/10.1007/978-3-031-78283-1_5

- The drifting of continents during geological time periods has been a cause for variation in climate for tens of millions of years on a particular continent. The continent/s drifting towards poles experience glaciation and continent/s drifting towards equator become warmer.
- Short-term variations in the atmospheric circulation patterns and astronomical factors also play role in the amount of solar radiation reaching the earth's surface that in turn influences the glaciation.
- The fluctuations in the output of energy sourced from the sun also influences the conditions for glaciation on earth's surface.
- Large amount of dust emitted with volcanic eruptions enhances the atmospheric solar reflectance which results in the drop in global temperatures causing the cooling of earth.

Glacier, covering approximately 11% of the earth's land surface, is a mass of ice and snow that moves down the slope of a hill/mountain or spreads outward on a land surface or continent and holds about 75% of the fresh water. Glacier is a dynamic mass of ice that moves under the influence of gravity (slope) and temperature conditions from higher to lower altitudes in mountains. The formation of a glacier begins with snowfall comprising almost 90% of air. Subsequent snowfalls build overburden and flakes of snow are packed into grain that becomes coarser and larger in size. This conversion of snow to ice depends on temperature conditions. The low summer temperature acts a necessary condition for the formation of a glacier ice as low temperature disallows melting of all snow. The snow that survives melting in summer and undergoes the process of transforming into solid glacier ice is called firn. The survived snow of last/previous year that is partially compacted and re-crystallized forms névé. The firn metamorphoses into solid glacier ice wherein air in the passages between interconnecting grains is completely sealed due to compression and bubbles of air are formed. This process of transformation makes firn denser than snow and ice denser than firn. The accumulated ice and snow, over a period, begins to slide/ move down the slope of a mountain or spread outward on the land surface, assuming the name glacier. The layers of snow that have changed to glacier ice by compression are distinct alternate of fine-grained and coarse-grained ice. The formation of glacier, therefore, is a gradual transformation of fresh snow to glacier ice through the process of firn/névé. Regular crystalline structure consisting of a single oxygen atom covalently bonded to two hydrogen atoms (H–O–H) forms the glacier ice. The ice frozen at atmospheric pressure is approximately 8.3% less dense than liquid water. The density of ice is 0.9167 g/cm$^3$ whereas water has a density of 0.9998 g/cm$^3$ at 0 °C. The frozen part of the earth, either land or ocean, is called cryosphere.

## 5.1 Morphology of Glacier

The body of glacier is divided into two zones, accumulation and ablation zones. The body of glacier where the snow accumulates and transforms into ice is the accumulation zone, whereas ablation zone is the one where snow is under constant melting. The accumulation zone grows in thickness and area and the ablation zone loses thickness and area over a period. The increase or decrease in thickness and area of the zones determines the health of a glacier. In case the thickness and area of the accumulation zone increases over time, the glacier is considered to have attained positive mass balance and indicates at the healthy state of the glacier. However, in case the thickness and area of accumulation zone of a glacier decreases, the glacier is considered to have negative mass balance and indicates an unhealthy state of the glacier. This gain/loss of mass of a glacier is used as a proxy to decipher climate conditions of the region. The zone that lies between the areas of accumulation and ablation, is known as the equilibrium line where there is neither gain nor loss in thickness and area of the glacier body. The equilibrium line fluctuates with increase or decrease in accumulation/ablation zones. The equilibrium line shifts down in elevation in case there is increase in the accumulation zone and decrease in the ablation zone. The equilibrium line shifts up the elevation where there is decrease in the accumulation area. The equilibrium line, therefore, is known as Equilibrium Line Altitude (ELA).

The glaciers are classified on the basis of various parameters, such as altitude and latitude. The alpine or mountain or valley glacier is formed at the altitudes in mountain regions of the world irrespective of the latitudes. The alpine or mountain or valley glacier is mostly longitudinal glacier extending from cirque to snout and fed by tributary glacier/s that is/are transverse to the longitudinal profile of the valley. The tributary glacier is, therefore, known as a transverse glacier. The degeneration of valley glacier, due to climate factors, facilitates the withdrawal of the glacier to higher altitudes of the mountain to occupy troughs and hollows, known as niche glacier. The transverse or tributary glacier may withdraw to elevation higher than the valley glacier and get disconnected with longitudinal glacier to form hanging glacier. The piedmont glacier occupies steep slopes of the valley and spreads out on a relatively flat slope with lobe-shaped snout. The mountains covered extensively with snow/ice can be source for more than one glacier that flows out from the extensive cover known as ice field. The ice field may cover the continents and is called continental glacier. The continental glacier preferably occurs at higher latitudes and is also termed as polar glacier.

The classification of the glacier is also based on the temperature profile. The cold glacier is the one where temperature gradient of a glacier is well below the pressure-melting point. The thin film of water formed because of pressure-melting is absent at the base or sole of the cold glacier. The cold glacier, therefore, has very negligible or no movement. The glacier with thin film of water at the base or sole is warm or temperate glacier and shows significant movement. Both types of glaciers, warm and cold, occur in Himalaya. The warm type of glacier is confined to watersheds of

Sikkim, Uttarakhand, Himachal Pradesh and Union Territory of Jammu & Kashmir in India. The cold type of glacier is found in the cold desert of Union Territory of Ladakh and adjoining areas.

The presence or absence of debris cover on the surface of glacier body also serves a parameter to classify the glacier. The body of glacier covered with debris is known as a dark or dirty or debris-covered glacier whereas the glacier devoid of any debris on the surface of glacier body is known as clean or debris-free glacier. Most of the glaciers in the Himalaya are dirty or dark glaciers.

The surge glacier is characterized with rapid advance of snout of glacier due to the transfer of mass (snow/ice) from accumulation to ablation zone in short period of time. Group of surging glaciers is present in Karakoram Himalaya. The rapid transfer of mass (snow/ice) from accumulation to ablation zone of the glacier occupies wide open space available in foot of the valley to form expanded foot snout of surge glacier. A narrow- or needle-shaped snout is observed in a glacier which is rapidly and constantly advancing. The end or terminal part of a glacier body, known as snout, can be of various shapes and sizes. The broad snout can occupy the entire width of the valley in case of large-sized glacier such as Siachen glacier, whereas lobe-shaped snout resembles an arc convex shape and is wider than the rest of the body of glacier. A cave in the ice/snow at the snout of a glacier, known as ice cave, is formed allowing the melt water to flow out in the form of river from the cave. The ice cave known as Gaumukh, formed at the snout of Gangotri glacier, is a noteworthy example.

Glacier moves from higher to lower elevations under the action of gravitational force developed by virtue of their own weight. Movement of glacier is further facilitated by either independently or collectively of the following processes.

a. *Internal deformation*: The creep movement occurs in a glacier due to the sliding of ice crystals that past each other leading to internal deformation. The rate of deformation is high in the basal part of glacier because of the development of high pressure, whereas upper part of the glacier is fragile and develops cracks or crevasses on the surface of a glacier. This type of movement occurs both in polar and temperate glaciers.
b. *Basal sliding*: A thin film of water produced at the base due to the pressure-melting at sole of the glacier reduces friction between moving glacier body and bedrocks enhancing the rate of movement of the glacier body. The extremely low temperatures in polar regions do not support the movement of glacier by basal sliding process. The temperature of ice mass plays a pivotal role in the movement and morphology of the glacier. The temperature of glacier ice, approximately at pressure-melting point throughout the thickness except the uppermost few metres, may decrease because of increasing pressure due to overburden. The geothermal heat generated due to the movement of glacier reaches the base producing meltwater in varying quantities at the base facilitating erosion, transport and sedimentation due to basal sliding.
c. *Bed deformation*: The presence of soft-sediments or weak rocks as bedrocks of glacier valley adds to the movement of glacier. The concentration of debris in the basal ice can influence the rheology of ice and friction of the bed, which

both impact the ice flow. Polyphase deformation indicating water-escape structures are typically related to late-stage ductile deformation. Evidence of brittle deformation is indicative of shear zones formed during later phases of syn- and post-depositional deformation.

The differential movement of the glacier body develops crevasse or crack on the surface of the glacier. The arc-shaped crack or crevasse at the head of glacier body separating the cirque from the valley walls is known as bergschrund; the crevasse which is more or less parallel to the length of glacier body is known as longitudinal crevasse; the crevasse which is more or less transverse to glacier body is known as transverse crevasse; and crevasse running more or less in zig-zag form is known as chevron crevasse. The crevasse that develops in the front/snout and on the sides / lateral margins of a glacier, are known as radial and marginal crevasse, respectively. Splaying crevasses develop when ice flow decelerates downslope, glacier bed flattens or rises, and where the valley widens on the inside of a bend and at the snout.

## 5.2 The Processes

The dynamics of glacier action involves the application of force over the land surface having gradient in a geomorphological system. The temperature of glacier ice mass plays a pivotal role in the movement and morphology of the glacier. The geothermal heat created by the movement of glacier reaches the base of the glacier to produce varying quantity of meltwater that facilitates sliding of the glacier body to generate the processes of erosion, transport and sedimentation. Climate cycles of glacial and interglacial periods influence the erosion rates with advancing glaciers remobilizing and transporting paraglacial or nonglacial sediments and retreating glaciers stimulate transport and erosion.

### *5.2.1 Glacier Erosion Process*

The landforms produced by virtue of erosion, transportation and deposition by the glacier varies from the head to the base of the glacier basin. The prime factors deciding the variable landforms within the glacier basin are distribution of glacier body at different altitudes, variation in lithology of entire basin, variable thickness of ice of the glacier body and fluctuations in the mass balance. The process of erosion by glacier, therefore, is a factor of thickness of glacier ice/snow, rate of movement of glacier snow/ice and residential time of the glacier body to act upon the land surface. The striae/striations and chatter marks on bedrock surfaces indicate the effect of glacier abrasion, whereas effects of high-pressure meltwater are indicated by scour marks. Glacier abrasion has a dominant role in producing most distinctively glacial features, smooth surfaces with striae and glacial polish (Siman-Tov et al. 2017), sometimes marked with crescentic gouges and crescentic fractures related to the episodic movement of large clasts. The abrasion at the sole of glacier is primarily by clasts carried in basal ice contacting the bed that are pressed downward by vertical ice flow balancing basal melting and lateral extension, and by their weight more than buoyancy provided by the ice (Alley et al. 2019 and references herein). The clasts move forward in the sliding ice, but are partially retarded by friction against the bed, which is in turn overcome by drag as ice flows horizontally around the clasts via enhanced creep and regelation. The subglacial fluvial action creates tunnel valleys (tunnel channels) in both sediment and bedrock, though not dominant in terrain (Kearsey et al. 2018), by vigorous and relatively extensive subglacial fluvial erosion (Beaud et al. 2018).

In Liddar valley, Kashmir Himalaya a major part of U-shaped valley and asymmetrical cross profile consisting steep clear-cut water face slope with a regional step-like structure along the valley wall are carved out by glaciers modifying the pre-existing valley spurs. Pleistocene glaciation typically started in mountainous terrain on steep slopes, and glacial-geomorphic processes led to subglacial erosion or proglacial deposition faster than in the proglacial environment, and consequently tended to shape the land surface into over-deepening that reduces the transport capacity of subglacial streams (Alley et al. 2019). Glacier erosion is guided by the geological structure of glacier valley/basin. The intensity of glacier erosion is more in the middle segment of the valley/basin and erosion in the upper and lower segments is less intense. The intense erosion in middle segment is because the movement of glacier ice is, more or less, parallel to the valley/basin segment. The erosion landforms are primarily formed by the process of glacier abrasion and plucking of rocks of the valley/basin. The plucking, a major erosion activity, weakens the bedrock beneath the glacier body and valley wall rocks. The well-jointed rocks with favourable spacing between joints / fractures is most suitable for the activity of plucking by glacier ice. The fractures in the rocks could be the result of overburden weight or basal stress near the margins of cavities that may intersect with structural weaknesses, such as bedding planes and joints affecting the overall rate of quarrying (Alley et al. 2019 and references herein).

The basal erosion by glaciers and frost heaving contributes to the steepening of valley walls which is facilitated by landslides, rock falls and avalanches leading to the genesis of loose rock material (debris) on the surface of glacier (Delaney and Anderson 2022 and reference herein). Active tectonics in High Mountain Asia, Alaska, Andes, etc. further add to the debris (Herreid and Pellicciotti 2020) which in turn effect surface to insulate glacier ablation zones from melt (Anderson et al. 2021 and references herein). The insulating effects lead to the accumulation of till under debris-covered ice is because (a) debris cover reduces meltwater delivery to the bed which reduces the rate of increase of subglacial water discharge and sediment transport capacity down glacier, and (b) debris cover also lowers glacier surface slopes which reduces sediment transport capacity relative to sediment availability (Delaney and Anderson 2022). The tendency of high number of glaciers to surge in Karakoram is subject to the formation of subglacial till, a result of connection between supraglacial and subglacial sediment environment (Bhambri et al. 2017).

Though, glacial erosion is predominantly at or above ELA, yet the geothermal heat from bedrock underlying glacier can significantly affect the thermal regime of the glacier base without any changes in surface conditions and/or ELA (Lai and Anders 2020). Besides, precipitation if reaches the bed of glacier, can promote basal sliding, glacial erosion and evacuation of sediments. Limited sliding at the base and meltwater, in the case of cold-based glaciers, results into minimal erosion compared to warm-based glaciers (Cook et al. 2020) suggesting that glacial erosion is a factor of some optimal climatic conditions (Mariotti et al. 2021).

It is evident from recent studies that erosion by glacier is governed by the amount of ice thickness, stress levels at the base of glacier body, rate of velocity of glacier ice at the base, content of the debris at the base of glacier and the temperature conditions at the erosive interface. The basal sliding process because of steep slopes accounts for 90% of total movement of glaciers in Himalaya. The load carried by valley glaciers of Himalaya is bimodal comprising large angular fragments plucked from the base as well as walls of the glacier valley and fine debris cover. Kashmir Himalaya glacial landscape, particularly in Liddar valley, includes bench cuts or sharp break in slope at the junction of cirque and rock basin which is produced by selective erosion of highly jointed/fractured bedrock by glacier. The break in slope produces a step-like structures having extreme gradient index value and subsequently modified by glacier abrasion.

### 5.2.2 *Glacier Transport Process*

The nature of mountain slope of the basin and velocity of glacier ice mass with sediment load governs the transport by the glacier. The sediment load on the glacier ice surface comprises the rock material that falls from adjoining valley walls, cirque and scoured valley bed and is termed supraglacial material. The boulder trains and isolated erratic boulders strewn across the glaciated landscape are evidence to the transport of rock material by the glacier. Transport pathways in the valley glacier are in response to gravity, differential solar radiation and fluctuation in mass balance of the glacier. During movement, glacier acquires subglacial material that is produced by the erosion of valley/basin floor, supraglacial material supplied from the walls and head of the valley/basin and pre-glacial rock debris from the valley/basin head. The carriage of large erratic, perched boulders and large-sized boulders as part of englacial process indicates the capacity of glacier to transport immense heavy sediment load. The preferred orientation of long axis of pebbles/coarse material and fabrics of the fine sediments transported by glacier, and striation marks on bedrock indicates the direction of the movement of glacier (Fig. 5.1a, b).

The evacuation of products of bedrock erosion from glaciated catchments is subject to the ease with which sediment can be remobilized and transferred by ice and meltwater (Cavalli et al. 2019, Perolo et al. 2019). The relationship between glaciation (glacier cover, sliding velocity, glacier thermal regime, glacier hydrology, subglacial and proglacial sediment cascade) and erosion rates (Cook et al. 2020, Perolo et al.

**Fig. 5.1** **a** Sh. V.K. Raina (Former Deputy Director General, Geological Survey of India) sitting on the striated boulder near the snout of Machoi glacier **b** striated and polished walls in the vicinity of the snout of Siachen glacier, Nubra valley, **c** rilled lateral moraine of Durung Drung glacier, Zanskar, Ladakh, **d** curvilinear lateral moraine of Kangriz glacier, Suru, Ladakh

2019) determines the transfer of sediments stored below and at margins of the glacier. The increase in export of sediments from glaciers during high-pressure events, also called 'spring events', is the reorganization of subglacial network allowing enhanced sediment export followed by decay in sediment yield (Lewington et al. 2020) and is likely to occur in summer too, due to high-amplitude daily variation in discharge (Davison et al. 2019). The change in the yield of sediments from the glacier is dependent on the variation in the mass balance of a glacier causing periods of intermittent enhanced and reduced sediment yield, known as sediment 'pulses' and 'hiatuses' (Alley et al. 2019). In a situation where meltwater reaches the bedrock, through crevasses and moulins, the subglacial drainage is dominant in flushing the bedrock eroded material from some parts and pre-glacial sediments from other part (Alley et al. 2019), whereas where subglacial drainage is inefficient, the bed is blanketed with thick protective layer of sediments reducing the erosion and evacuation. The ice pressure, water pressure and thermal regime at the glacier bed dictates the distribution and spatial reorganization of the subglacial drainage network, from time to time, depending upon the inputs of meltwater and sediment and ice flow dynamics, influencing the efficiency of sediment transport (Delaney et al. 2018, Perolo et al. 2019).

### 5.2.3 Glacier Deposition Process

The deposition by glacier depends upon the ice temperature, rate of velocity of glacier ice and nature of subglacial sediments. The sediments on the surface of glacier and carried by the glacier erosion of basin together form distinct landforms when deposited. The landform features such as moraines, drumlins, eskers, glacial lake sediments and glaciomarine sediment, ubiquitous to glaciated terrain, are result of deposition. The mechanics of deposition by glacier is associated to the advance and retreat of glacier in the valley/basin. The depositional landforms, formed as a result of the advance/retreat, provide information on the history of Pleistocene glaciation. However, availability of debris (subglacial and supraglacial) and glacier dynamics decides for the timescale in the building of frontal moraines (Bernard et al. 2020). The subglacial dynamics is perplexing as increase in water pressure reduces the contact area between glacier sole ice and bed due to the "hydraulic jacking" of the overlying ice, thereby increasing the rate of basal motion (Davison et al. 2019 and reference herein). The subglacial sediment transport is function of water discharge coupled with ice dynamics as pressurized water flow increases the velocity and exerts shear stress on the sediments to mobilize (Delaney et al. 2018), that are abundant in subglacial environment compared to fluvial environment due to deformation and englacial transport and deposition of till into subglacial conduits.

Different climates govern the pattern of glacial erosion to give rise to distinct landforms. Over deepening and high relief of valley is noticed in cold and dry climate conditions where glacial erosion rate exceeds the bedrock uplift rate and limited amount of erosion occurs in higher elevations in cold-based glaciers preserving the pre-glacial landforms (Lai and Anders 2021). On the other hand, the topography is reshaped due to high rate of erosion in the higher reaches under warm and wet climate conditions.

The glaciofluvial sediments and gravel emerging out from beneath the glacier body is deposited linearly in the form of conduits for several metres to tens of metres in height and width giving rise to eskers or esker field. There can be still more complex eskers formed under proglacial, supraglacial and subglacial environments (Storrar et al. 2019). The magnitude of stress enhancement around an obstacle decides for the length of subglacial bedforms, flutes or drumlins (Fuller and Murray 2020). Drumlins have evolved from an instability of traverse form (as in Rogen moraine) explained via the interaction of ice flow and till flow (Fowler 2000). Off-loading of supraglacial material gives rise to lateral moraine ridge, though with contribution from englacial and subglacial material too. The annual pushing of sediment at the margin forms the ice-marginal moraines whereas, glacier surface with sufficient debris cover to retard ablation produces ablation moraines (Graham and Midgley 2000). Melting of ice-cores lead to a landscape of chaotic mounds that are subjected to repeated process of sedimentation.

Both, brittle and ductile, phases of deformation are possible in glacial sediments with sufficient intersection of both phases during the process of development of landforms (Menzies 2000). Shear and low viscosity due to higher temperature at the bed of glacier act as important drivers in the deformation to the extent of generating folds on shallow undulatory layers (Souchez et al. 2000).

Repeated glacial and interglacial cycles of Quaternary glaciation are mainly responsible in sculpting the present-day landscape of Himalaya (Shukla et al. 2019; Garg et al. 2019). The mountain ranges of Kashmir valley were glaciated during the late Quaternary period (Murtaza 2019) evident from the study of palaeo-landforms (glacial moraines, loess-paleosol sequences and lacustrine sediments) (Shah et al. 2020). The studies (Dar et al. 2017; Rashid et al. 2017) further revealed three major glacial advances during late Quaternary period in Thajwas glacier and Kolahoi glacier valleys. Estimation of palaeo-thickness and reconstruction of palaeo-extent of glaciers in mountains of Kashmir valley is recorded in geo-archives in the form of striations, grooves, lateral moraines, outwash plains, etc. (Murtaza et al. 2021 and references herein). Besides the direction of flow, the striations and grooves also indicate the nature of coarse material and abrasive power of the glaciers in Kashmir Himalaya. The maximum height of lateral moraines and trimlines provided an estimated thickness of glaciers to approximately 200 m during the Quaternary period in Kashmir Himalaya with an extent of ~ 7 km below the present snout position.

Study of palaeo-landforms from Suru valley (Kumar et al. 2021a) identify first advance of compound valley glaciers during 33–23 ka from Pensi-La (~ 4454 m amsl) to near Achambur (~ 3200 m amsl), an extent of ~ 70 km downstream with an average thickness of ~ 173 m, followed by second advance during 17–15 ka with an average thickness of ~ 150 m, third advance during 13–11 ka with an average thickness of ~ 94 m, fourth advance during 10–7 ka with an average thickness of glacier to ~ 86 m and last advance during 2.8–2.3 ka with average thickness to ~ 75 m. Similarly, the glacial landforms unravel the glacial history of Lahaul Himalaya dating back to ~ 186 ka (Das et al. 2023). Four major glacial advance stages are unfolded from the studies—Chandra glacial stage (~152 ka), Batal glacial stage (~43 ka), Dali glacial stage (~20 ka) and Mayar advance (LIA) -suggesting change from an extended single ice mass ~ 10 km down-valley to large valley glaciers receding further to cirque and small valley glaciers.

Three category of lateral moraines, active, recessional and static, were identified through systematic study of moraine complex in the glacial landscape of Satopanth glacier valley, Garhwal Himalaya (Sharma et al. 2018a). The study further suggested that medial and supraglacial moraines with debris cover are confined to the upper part of mid-valley. Dead ice mounds are documented near the snout of Chorabari glacier formed by the compaction of subglacial loose material in glacier ice that were detached from the glacier body due to the formation of basal transverse crevasses (Sharma et al. 2018a). The Miyar glacier (Lahaul Himalaya) showed a retreat of 471 m between the end of LIA (1850AD) and 1965 at an average rate of $4\text{my}^{-1}$, 429 m between 1965 and 2019 at an average rate of $7.9\text{my}^{-1}$ and increase in average rate of retreat to $11.5\text{my}^{-1}$ between 2013 and 2019 (Deswal et al. 2023), whereas Kumar and Krishna (2021) have reported an overall retreat of 1205 m between 1962 and 2017. The Pimu glacier (Lahaul Himalaya) shows a retreat of ~ 2.2 km since 1850AD at an average rate of $13\text{my}^{-1}$ with highest rate of retreat between 1980 and 2002 ($10.5\ \text{my}^{-1}$) which has slowed down to $\sim 5.6\text{my}^{-1}$ in recent times. Menthosa glacier (Lahaul Himalaya) retreated ~1235 m between 1850 and 1965 at an average rate of 10.7 $my^{-1}$ and slowed down to 5.1 $my^{-1}$ between 1965 and 2002 with a glacier retreat of ~366 m but accelerated between 2002 and 2013 at an average rate of 7.3 $my^{-1}$ (Deswal et al. 2023). Seven stages of glacial advance are reported from Dhauliganga basin (Central Himalaya) with oldest dated between 81 to 41 ka (MIS4 to mid MIS3) and youngest between 2 and 1 ka (late Holocene) (Ali et al. 2022). The expansion of advance is restricted after second glacial advance dated between 26 ± 5 ka and 17 ± 2 ka (coeval with LGM) in the basin.

The study of erosional and depositional landforms of Alaknanda Basin (Central Himalaya) indicates three major glacial advances. Studies suggest the extent of Satopanth, Bhagirath Kharak and Bhagnyu glaciers of the basin at a distance of 8.4 ± 1.1 km, 7.8 ± 1 km and 8.4 ± 1.1 km from the present-day snout of glaciers, respectively during the Stage I of the advance (Remya et al. 2024), while the Stage II moraines are 5.1 ± 0.71 km, 4.2 ± 0.58 km and 5 ± 0.7 km away, and Stage III moraines are 1.9 ± 0.26 km, 1.3 ± 0.18 km and 2.8 ± 0.39 km away, respectively.

## 5.3 Glacier Geomorphology and Landforms

Glacier geomorphology facilitates to understand and interpret the inter-relationship between topography, morphology of glaciated landscapes and associated sediments. The study of landforms helps to understand the myriads of processes that occur within several sub-environments of glaciated terrain. The landforms are result of inter-relationship between several sub-environments of the glaciated terrain, such as supraglacial, englacial, subglacial, terrestrial proglacial, sub-aquatic proglacial. The delivery system of sediments in sub-environments play role in the development of glacier landscapes or landforms. Micro-morphology and micro-sedimentology of glacier provides interpretations on processes and rheological changes in subglacial sediments (Woodward et al. 2018, Phillips et al. 2018).

The present-day spectacular topography of northwest Himalaya is a result of activities by glacier/s in different watersheds in the recent past. The accumulation and ablation zones are separated by geomorphic threshold in the glacier. The transfer of threshold indicates basal slippage of the glacier and causes sanding action at the base of the glacier due to the transfer of energy by either advance or retreat of the glacier that initiates the glacier process (Singh et al. 2020).

The landforms in the Himalaya have been produced by the glacier processes in and around glacier valleys covered with snow and ice since Pleistocene-Holocene times. However, the storage of sediments below and at the margins of a glacier and time required for the transfer of sediments across the landscape is a matter of debate. The assumption of glacial landform genesis is a product of intricate relationship between the drivers such as glacier cover, sliding velocity, glacier thermal regime, glacier hydrology, subglacial sediments, proglacial sediments and erosion rates (Cook et al. 2020; Perolo et al. 2019). The assumption is further complicated due to high variability in sediment yield through glacial cycles generating phases of glacier advance, retreat and re-advance that affects changes in ice velocity, erosion rates, ice and melt-water transport capacity and glacially conditioned sediment accessibility at the bed (Antoniazza and Lane 2021).

Taking into account that present-day landscape of the Himalaya is primarily sculpted by the repetitive cycles of glacial and interglacial during the Quaternary period (Shukla and Garg 2019; Garg et al. 2019), geomorphological survey of glacier basins in Indian part of Himalaya was undertaken to study snout positions to measure the rate of movement, thickness of ice and glacial cover area through geological timescale (Lin et al. 2017; Zhou et al. 2017; Bolch et al. 2017; Gao et al. 2020 to quote a few). Significant progress has been made in recent years by the application of optically stimulated luminescence (OSL) and cosmogenic radionuclide (CRN) methods allowing the dating of glacier sediments and landforms, particularly in the western and Karakorum Himalaya of India. The field information supplemented with chronometric data on the extent of glaciers in the past is significant to understand the flexibility in climate vis-à-vis geomorphic development of landscape in Himalaya (Dar et al. 2017). However, the inaccessible rugged terrain and security constraints have served challenge in Himalaya creating a wide knowledge gap. In spite of a few glacial-geomorphologic studies in the Himalaya highlighting the importance of glacier geomorphology in the reconstruction of the Quaternary glaciation and climate variability (Dubey et al. 2019; Singh et al. 2020), there are several sub-ranges in Himalaya which are unstudied till date due to the aforesaid challenges.

### 5.3.1 *Liddar Glacier Valley*

Liddar River is one of the main tributaries of River Jhelum and commences from the base of the two main ice fields, the Kolahoi (Kolahoi-Hubungan ridge) and the Shishram (Zamtirnag-cirque of Kunyirhayan peak). A major part of the U-shaped trough carved in the east and west Liddar valley at an altitude of 3600 m amsl is buried under ice. The striated rock surfaces near Sheshnag lake, above roche-montannees, reveal that the thickness of ice responsible for major glaciation was not less than 250 m. Geomorphological reconnaissance of the valley up to Ganshibal (2000 m amsl) documents preservation of abundant drift material in the form of moraines and till, and erosional features in the form of cirque, nivation hollows, scoured bedrock, roche moutonnée, kames and eskers. The presence of polished and striated rock surfaces on roche- moutonnée and whaleback at mean elevation of 2200, 2700 and 3200 m amsl in the valley/basin indicate three stages of glacial advance. Morphometric analysis of cirques, particularly related to length-height and length-breadth ratio and study of clustering pattern of cirques to assess the shape and duration of the occupation of ice, shows that most of the large cirques and cirque lakes in east Liddar are cylindrical finger type indicating long occupation of ice, compared to rectangular type in west Liddar indicating thin ice cover. Clustering of cirques at three different elevation zones of 2200–2700, 2800–3500 and 3650 m amsl and above indicate three stages of glacial advance in the valley. The cirque levels are in sync with bench cuts and glacial scoops formed on the valley walls. Nearly 40% of the glaciated area of Liddar valley is covered with depositional features with predominance of moraines representing 70% of the depositional landforms (Fig. 5.2). Lateral moraines, aligned parallel to the valley margins and slightly arcuate, are the most common feature. The moraine joins their counterparts in hanging valley at acute or right angle. The moraines are divided into high, medium high and low in terms of their location at 100–200 m, 60–100 m and 25–50 m from the valley bottom, respectively. Length and breadth of moraines vary from a few metres to 7 kms and 50 m to 2 kms, respectively. The lateral moraine ridges in east Liddar extend from Ganshibal (2000 m amsl) to Pahalgam (2200 m amsl), Aru (2600 m amsl) and Satlanjan (3500 m amsl) to Kolahaiganj (3750 m amsl). In west Liddar the ridges extend from Pahalgam to Chandanwari (2700 m amsl), Zagpal (3650 m amsl) and Shishram (3750 m amsl) (Fig. 2.7). The low moraine ridges, curved in form, extend from Sheshnag to Vaojan and are superimposed with fresh truncated moraines near Shishram, Kolahaiganj and Dudnag cirque in east Liddar. The medium high lateral moraines are present at Zaijpal (east Liddar) and Kalwan-Lidderwat (west Liddar). The older lateral moraine at 120 m and younger lateral moraine at 60–100 m above valley bottom, respectively suggest two stages of glacier advance in the glacier valley. The presence of lobate recessional moraine with furrow structure, kettle hole, kame mounds and eskers on distal slope in northwest of Saijpal-Vaojan at 3750 m amsl in east Liddar and at Satlanjan in west Liddar suggest slow ice advance due to resistant basin edges causing post-depositional modification of medial moraine. The similarity in the trends of moraines at Zaijpal in east Liddar and Satlanjan in west Liddar points at the time contemporaneity of formation of moraines. The pebbles in the till of lateral moraine show preferred orientation parallel to the crest of moraine. The uniformity in lithology of large coarse fraction of high, medium high and low moraines is attributed to a single glacial advance but variation in the frequency of boulders and weathering of the soil suggest difference in their relative ages. The outer moraine ridge of medium high moraine near Lidderwat and Zojipal have been obliterated by the second glacier advance. The third advance of glacier removed all the earlier evidence except that of outer medium high morainic ridge near Saijpal and Liddarwat and the advance extended to Ganshibal that is indicated by huge terminal moraine of 17 kyr in age. The location of three distinct end moraines in Liddar valley at elevations of 2000–2150, 2600–2800 and 3200–3400 m amsl near Ganshibal, Kalwan-Chandanwari and Satlanjan-Saijpal indicate retreat of glacier advance in three phases. The end moraines near Satlanjan and Sheshnag are approximately 1 km long, 500 m wide and 25 m to 30 m above the river bed. The push moraines in east and west Liddar are well preserved at Aru and Chandanwari. The three unpaired terrace levels located between 1930 and 2200 m amsl are the manifestation of three glacial advance and retreat process that occurred in Pleistocene-Holocene period. The three stages of advance/retreat occurred at 17–10 kyr (warm temperate climate), 10 kyr (cool climate), 8kyr (warm humid climate) and 6 kyr BP (cool climate) as suggested by the analysis of pollen and lichen.

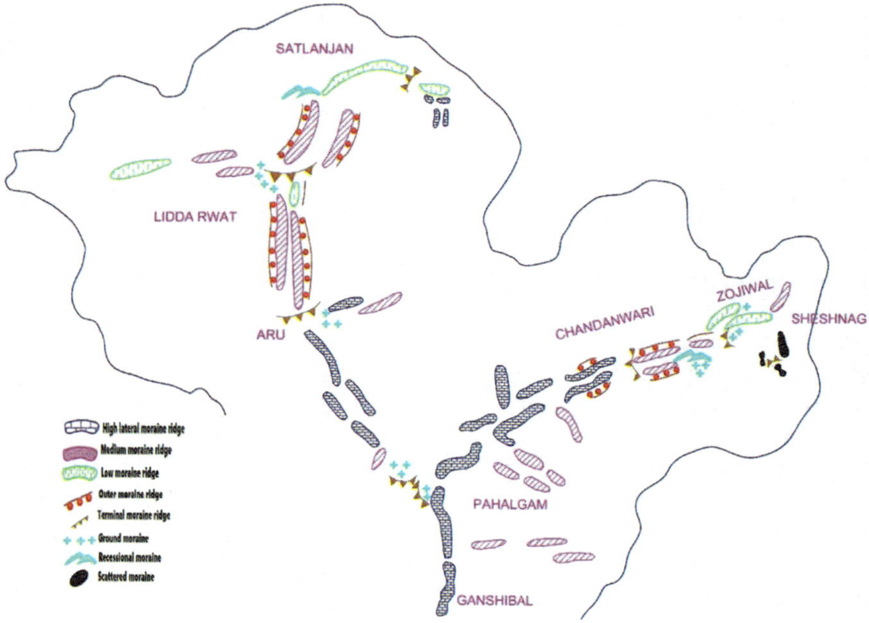


**Fig. 5.2** Geomorphological map of Liddar valley, Kashmir

### 5.3.2 *Baspa Glacier Valley*

The tributary glaciers, such as Kupa, Khargola, Dungdengsee, Gor Garang, Rakcham, Mastrang, Karu, Naradu Garang and Jorya Garang, descending from high mountain peaks of Whoming (5215 m amsl), Khomsung (5700 m amsl) and Khimsung (5600 m amsl) preserve evidence of glaciation in the form of cirque, nivation hollows, scoured bedrock, roche-moutonnée, kames, eskers and moraines. The Baspa Bamak and Naradu glaciers (Fig. 5.3a) have snouts covered with supraglacial material and with irregular outline. The glaciers have four treads punctuated by steps. The convergence of two medial moraine loops has developed recessional moraine breached by Baspa River. Compact moraine comprising sandy-silt matrix and well-rounded boulders of phyllite, and quartzite are traced 700 m down the stream in both Baspa and Naradu Garang Rivers. Predominant presence of striation marks on valley walls of Naradu, Nagasti and Suti point at ~ 300 m thick glacier ice. Valley in valley profile is produced in main and tributary valleys of Sangla, Rakcham, Chitkul and Nagsti from source to confluence with U-shaped valley in the upper reaches and V-shaped notch at base. Three glacial stages and two advances are recognized in the glacier valleys of Baspa and the second advance being more extensive. Development of two high lateral moraine ridges at Sangla and Rakcham and recessional moraine with series of terminal moraines at Gainder *khad* in Baspa Bamak documents two glacial advances (Fig. 5.4). Three end moraine loops at the altitude of 2250 m amsl document the early

Baspa stage or first retreat (Kupa-Darcho) dated to 12,136-10,900 years BP, middle Baspa stage retreat (Rakcham-Mastrang) at the altitude of 3050 m amsl dated to 9800–4300years BP and late Baspa stage (Chitkul-Nagasti) at the altitude of 3400–4000 m amsl dated to 1800years BP. Two high lateral morainic ridges at the altitude of 3000–3500 m amsl and 4000–4500 m amsl indicate two glacial advances.

**Fig. 5.3** **a** Google Earth view and field photograph of the snout of Naradu glacier, Himachal Pradesh, **b** Google Earth and general field view of Durung Drung glacier, Zanskar, Ladakh

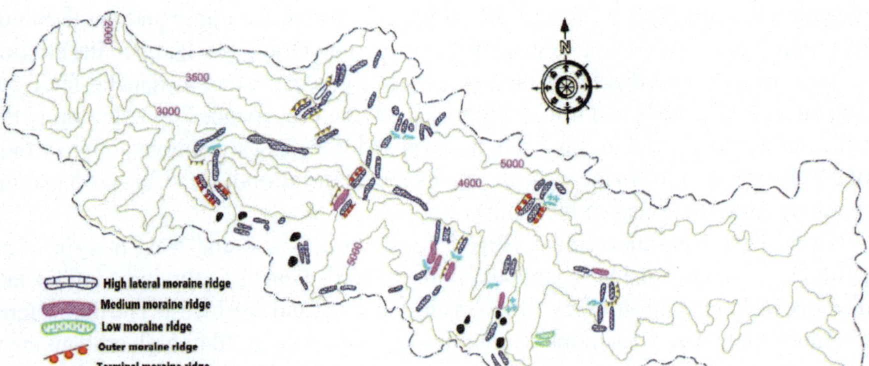


**Fig. 5.4** Geomorphological map of Baspa/Sangla valley, Himachal Pradesh

Three series of treads at Jorya Garang, Rakcham and Mangsa Garang in Baspa valley separated by 150–100 m high stairways are located at a distance of 7 km, 30 km and 54 km, respectively from source to confluence. The 50–60 m long, 10 m wide and 15–20 m high roche-moutonneé are well preserved at 3680 and 4300 m amsl in Baspa Bamak-Naradu Garang glacier valley. Part of rock-striated surfaces on roche-moutonnée at Rakcham (3200 m amsl near PWD guest house) and at Khomtsingh Garang (3660 m amsl) indicate two glacier advances in Baspa valley. Fusion of two medial moraines along with a series of terminal moraines explains the mechanism of development of recessional moraines in Naradu and Baspa Bamak valleys.

### *5.3.3 Durung Drung Glacier Valley*

The Durung Drung glacier (Fig. 5.3b) commences from the cyclopean cirque (one above another). The cirques are confined between 6090 and 5240 m amsl and at 5200 m amsl and terminate to an altitude 4000 m amsl. The intensity of erosion by glacier is dominant near the snout forming the eastern part of the valley and Pensi La segment forming the western part, and erosion in the middle and upper segment of the valley is less dominant. The occurrence of striations on valley walls, cirques and roche-moutonnée justify the intensity of erosion in the valley. The roche-mouronnée manifest coexistence of both the processes of polishing and quarrying at stoss side and lee side. The longitudinal profile of the Durung Drung valley reveals that floor of the valley is a broad stairway with three series of treads in the middle section and one small tread near the headwall. The treads are separated by stairways at elevations of 4280, 4820 and 5280 m amsl, with the development of three icefalls. The tributary valleys join the main valley at a slope of 35° to 40°. The hanging valleys are convex in form at the middle and slightly concave at the valley head. Overall, the profile is less steep in the middle part than the upper and the terminus part of the valley. Abundant debris covers the Durung Drung glacier up to the altitude of 4040 m amsl and thereafter the glacier surface is clean and avalanche-fed type. The erratics of granite and migmatite rocks, having provenance 20–30 km up in the valley to the south of Zanskar Fault, are exposed in Pensi La segment at the altitude of 4400–4000 m amsl in the west and near snout at the altitude of 4120–4165 m amsl in the northeast segment of the valley.

Large lateral moraine ridges (Fig. 5.1c, d) on the east and west margin at an altitude of 4480 m amsl have sinuous course towards Pensi La moraine complex at a distance of 15 km and altitude of 4080 m amsl in north and northwest. The high lateral moraines in the Pensi La moraine complex are preserved as 70 m high embankment on west margin of the valley wall that gradually loses height to 45 m at an altitude of 4080 m amsl. The high lateral moraine ridges near the mouth of Pensi La (Fig. 5.5) has developed series of compressed crested moraine ridges having height between 15 and 25 m and width between 70 and 100 m. The medium high lateral moraine having a basal width between 60 and 80 m, are well preserved along the margins of the valley between an altitude of 4690 and 4480 m amsl and run for a distance of 5kms. Two well-developed medial moraines, one in the vicinity of present-day main glacier valley and the second in the Pensi La area, are preserved. Three medial moraines, first in the west at an altitude of 4620–4470 m amsl, second in the east at an altitude of 4560–4350 m amsl and third in the central part of glacier body at an altitude of 4420–4200 m amsl, are 6–7 kms in length. The third medial moraine in central part of the glacier body is arcuate in form. The medial moraines in Pensi La area, confined to an altitude of 4380 m, are fork-shaped in form having length of 1.5 km and height varying between 30 and 45 m, and arms align towards south and southwest. The dumping of the debris of the glacier in the form of terminal and latro-terminal moraines give rise to supraglacial moraine-dammed lakes. The post-depositional modification of medial moraine in upper Pensi La between 4450 and 4400 m amsl is an outcome of the shift in glacier body and sedimentation of suspended material of dumped icebergs.

Large number of fluted moraines in the form of linear furrows are 2.5 m in height in vicinity of 4160 m amsl and cover an area of 2 $km^2$ of the valley floor. Their surface is covered with the rounded boulders of locally available crystalline, sandstone, granite and phyllite rocks. The feature has rolling outline which is attributed to the squeezing of water-soaked debris in channel cavities beneath the ice. Ice-cored moraines are in the front part of the Durung Drung glacier valley, particularly in the central and east part having varied dimensions and shapes. The length varies from 150 to 70 m, height from 5 to 35 m and width from 15 to 35 m. The ice-cored moraines have turned into calmy mounds after the moraines melted.

The re-activation of Zanskar Fault in Pleistocene-early Holocene uplifted the southwest ranges of the region and brought change in the style of glaciation by shifting the extension of glacier from west to northeast. The clustering of cirques at

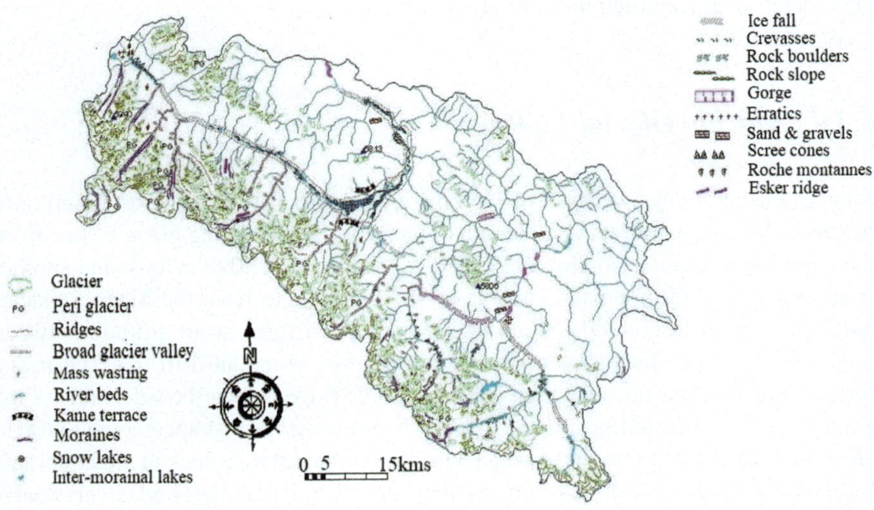


**Fig. 5.5** Geomorphological map of Suru-Zanskar valley

two different altitudes of 4950–5200 m amsl and 4500–4800 m amsl indicate two glacier advances, supported by the development of valley in valley profile particularly at an altitude of 4900 and 5300 m amsl. The Durung Drung glacier turns sharply to west near 4500 m amsl, where the width of valley floor increases from 1.9 to 2.5 km, upto an altitude of 4040 m amsl in northeast to build complex of landforms. Glacier Elevation Index (GEI), defined as difference between two states of accumulation area ratio (AAR) for glaciers—one is determined by averaging of the observed time-varying AAR*i* (where *i* is the year) and second is a hypothetical equilibrium AAR0 which is the accumulation–area ratio of the glacier if and when it is in mass-balance equilibrium (Dyurgerov et al. 2009)—shows increase from 4040 m amsl in northwest to 6020 m amsl in southeast agreeing well with climate snowline (Taylor and Mitchell 2000). Three terminal loops ranging in height from 5 to 10 m are formed at an altitude of 4180, 4200 and 4250 m amsl along with medial moraine on the basin floor in Pensi La area. The last stage of Durung Drung glaciation extended to an altitude of 4000 m amsl and turned in a sinuous course to enter Doda/Stod valley followed by the retreat. The elongated fluted moraine at an altitude of 4350 m amsl indicates rapid movement of the glacier over the humped ridge which acted as a barrier in smooth flow of the glacier and caused deposition of debris in the form of lateral moraines. During the Pensi La advance in early Holocene, the low relative relief in northeast towards present-day snout (4120 m amsl) enabled Durung Drung glacier to gradually shift and expand its foot to reach lower altitude of adjacent lowland plain in northeast after crossing over rock ridge. The activity of glacier resulted in the development of two moraines at Pensi La (4400 m amsl) and in the vicinity of present-day snout. The spreading of present-day Durung Drung glacier resulted in the development of broad foot terminal along with huge debris. The glacier reworked its debris by glacio-fluvial action because of the retreat and redeposited the sediments in the valley floor, both in Pansi La area and the main glacier valley giving rise to sandur outwash plains punctuated with ice-cored moraines (Fig. 5.6a, b).

### *5.3.4 Machoi Glacier Valley*

Machoi glacier shows vacation of 0.6 km$^2$ area in last fifty years substantiating to negligible change in the area and length of the snout. Two feeding glaciers, one from the main large cirque in southwest located at an altitude of 5090 m amsl and another from cirque at 5000 m amsl, converge at 4880 m amsl to form the Machoi glacier body. The snout is pointed tongue like (Fig. 5.7) in form at an altitude between 3680 and 3652 m amsl. The entire valley preserves abundant drift material in the form of moraines particularly high lateral moraine ridges along the side walls of the glacier valley from equilibrium line (4530–4515 m amsl) to V-shaped notch (3700 m amsl) and further down to the snout. The ridges of lateral moraines in slight arcuate form extend further to northeast up to Gumri *nallah* (= streamlet) and covers nearly two kilometres. The field evidence reveal that Machoi glacier was broad-footed and massive in early Holocene period. The embankments of moraine having slope of 15°

**Fig. 5.6** **a** Sandur outwash plain in the foreground of the snout of Durung Drung glacier with ice-cored moraines (conical mounds), **b** closer view of ice-cored moraine, **c** Steep scarp at the toe of alluvial cone in Shyok-Nubra valley

to 30° are between 60 and 100 m high from the valley bed. A series of three medium lateral moraines in the lower segment of the valley lie between the altitude from 3400 to 3700 m amsl on either side of Machoi *nallah* whereas subdued moraines occupy between the altitude from 3700 to 4100 m amsl in higher segment and coalesce with ridges of high lateral moraine at an angle of 25° (Fig. 5.8).

The trough head of the valley is carved out in volcanic rocks. The side walls of valley are very steep and wide; it passes over limestone and shales. The valley has developed convex profile, about 1.7 km downstream, with a hump at the top (4800 m amsl). Another glacier basin depression (cirque) is formed in the valley at an altitude of 4520 m amsl, where accumulation of snow and ice from main glacier as well as tributary glacier resulted in the formation of convex surface profile leading in the formation of another hump at an altitude of 4450 m amsl. Further down, the valley floor narrows at an altitude 4440 m amsl. A valley in valley profile is developed in the Machoi valley, particularly at an altitude between 4240 and 4440 m amsl indicating two stages of glaciation. The roche-moutonnée are situated in the vicinity of snout at 3700 and at 3450 m amsl.

**Fig. 5.7** Snout of Machoi glacier, Drass, Ladakh and shift in the position of snout between 2012 and 2014

## *5.3.5 Siachen Glacier Valley*

Siachen glacier, situated in Nubra valley, is the largest glacier in NW and Karakoram Himalaya (Fig. 5.9a) and oriented to southeast. The geomorphological features suggest young age to the glacier valley. From the snout of Siachen glacier to the confluence of Nubra River with Shyok River, the valley lacks convincing evidence of older moraines in ascertaining the palaeo-extent of the Siachen glacier. Most of the moraines are along the valley walls representing the remnants of tributary glaciers that descended from the small valleys in Saltoro range. Large number of scree deposits in Nubra valley merge with alluvium of the river. The scree cone deposits, 2–3 m high, formed at the confluence of scree with alluvium of Nubra River have steep scarps that confirm the tectonic rejuvenation of the valley (Fig. 5.6c) in geologically younger times, may be Holocene period. The activation of Karakoram Fault, running beneath the thick alluvium of Nubra valley, in Holocene/Recent period could be the primary cause for the formation of steep scarps. Close to the snout of Siachen glacier, there are no evidence of older moraines. However, 300–500 m down the snout of Siachen glacier, valley wall striations and eroded remnants of right lateral moraine sticking to the valley wall are the evidence to identify the mid-Holocene extent of the glacier. However, optical chronologies obtained on moraines indicate the oldest glacier advance dated to $60.4 \pm 5.2$ ka when ~ 700 m thick ice cover occupied entire Nubra valley extending beyond the confluence of Nubra-Shyok Rivers, followed by second major glacial advance dated to $30 \pm 2.5$ ka when the Siachen glacier reached up to the Nubra-Shyok confluence while the thickness of glacier ice is less (~ 300 m)

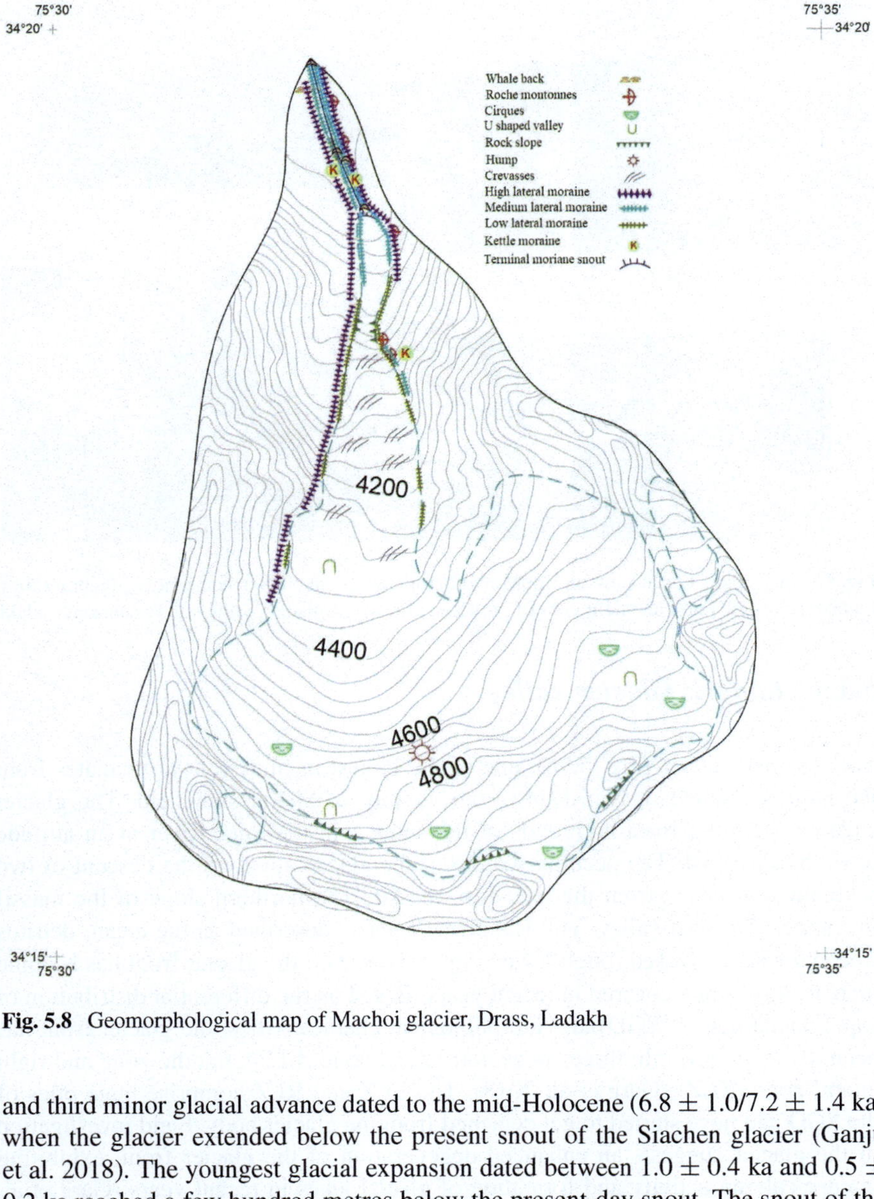


**Fig. 5.8** Geomorphological map of Machoi glacier, Drass, Ladakh

and third minor glacial advance dated to the mid-Holocene (6.8 ± 1.0/7.2 ± 1.4 ka) when the glacier extended below the present snout of the Siachen glacier (Ganju et al. 2018). The youngest glacial expansion dated between 1.0 ± 0.4 ka and 0.5 ± 0.2 ka reached a few hundred metres below the present-day snout. The snout of the glacier is zig-zag in outline from west to east and highly crevassed with deep longitudinal and transverse crevasses facilitating supra and englacial meltwater channels and frequent calving of the snout. The Nubra River originates from the east end of the snout.

**Fig. 5.9** **a** Google Earth view and field photograph of snout of Siachen glacier, Nubra valley, Ladakh **b** Google Earth and field view of the snout of Kangriz glacier, Suru valley, Zanskar, Ladakh

### *5.3.6 Kangriz Glacier Valley*

The Kangriz glacier (Fig. 5.9b), one of the largest in Suru valley, emanates from the large snowfield of Nun-Kun in the vicinity of Shafat snowfield. The glacier streams emerging from both ends of the snout join the Suru River at an altitude of ~ 3578 m amsl. The peculiar feature of the glacier involves the descent of two accumulation heads from the Nun-Kun massif. The northern slope of the massif overhangs the Suru valley. The Kangriz glacier is described as the clean, detritus free and with crevassed rough lower slopes. However, the glacier front has become increasingly debris covered in recent years. Based on the differential distribution of supraglacial debris (SGD) and morphological characteristics, the glacier front has been distinguished into three zones, left lateral zone (LLZ), middle zone and right lateral zone (RLZ) (Garg et al. 2018). The LLZ and RLZ comprise huge piles of the SGD and have started to get detached from the glacier body. Field investigation of the glacier suggests an enhanced degeneration of the glacier front, exhibiting intense calving activity and formation of glacial lakes in recent years (Garg et al. 2018), substantiating the observation made by the authors earlier on thinning (surface lowering) of the glacial surface over the last two decades. The findings on Kangriz/Parkachik glacier are consistent with those anomalously advancing glaciers found in the Karakoram. These glaciers tend to advance sporadically, particularly during the1990s and early 2000s. The anomalous behaviour of glaciers in the Karakoram was constrained to those glaciers with the highest levels of relief, and which cover extreme altitudinal ranges, the criteria met by Parkachik glacier which originates from the catchment basin at an altitude of ~ 6000 m amsl and have steep icefall along the tongue at the terminus in the Suru River at 3560 m amsl.

# Chapter 6
# Mountain Geo-Hazards of NW Himalaya

**Abstract** Largest number of fatalities due to disasters in 2020–2021 is reported from India (4743) as approximately 58.6% of Indian landmass is prone to earthquakes, 12% is prone to floods and river erosion and 68% of cultivable land is vulnerable to drought with risks of landslides and avalanches in mountainous regions of the country. The disasters in India are associated with winter, pre-monsoon, monsoon and post-monsoon seasons. The continuous uplift of Himalaya coupled with denudation, called tectonic aneurysm, triggers the natural disasters such as earthquakes, mass movements, avalanches, floods, glacial lake outburst floods (GLOFs), cloudburst, desertification, etc. The increase in number of glacial lakes to 5.9% in HMA (High Mountain Asia) between 1990 and 2015 brings both large opportunities and risks to the lives and livelihoods in the form of catastrophic events. The sudden outburst of blocked narrow and steep walled river valley by rock avalanches, slides and debris flows in mountainous regions can be the cause for devastating landslide lake outburst floods. Deforestation, slope cutting, overgrazing and tourism associated with heavy rainfall in mountainous region of India are key factors causing soil erosion that leads to landslides and rockfall. Increase in the frequency of ice/snow avalanches in Himalaya is attributed to glacier instability, a factor of disturbance in water-heat balance of glacier due to the fluctuating precipitation and continuous warming of the region. The forest fire events in India are largely caused by humans, such as grazers, sparks from vehicles, hiking or picnicking activities, shifting cultivation and burning litter (pine needles) for a flush of fresh grasses.

**Keywords** Tectonic aneurysm · GLOF · LLOF · Snow avalanche · Earthquake induced landslides · Forest fire · Emergency Events Database (EM-DAT) · Centre for Research on the Epidemiology of Disasters (CRED) · HiMAP (Hindu Kush Himalayan Monitoring and Assessment Programme) · RGI (Randolph Glacier Inventory) · HMA (High Mountain Asia)

The disasters in India are related to four distinct seasons—winter season (December, January and February), pre-monsoon or hot weather season (March, April and May), monsoon season (June to September) and post-monsoon season (October, November)



R. K. Ganjoo and M. N. Koul, *The Himalayan Glaciers, Climate Change and Society*,
https://doi.org/10.1007/978-3-031-78283-1_6

(Bhat and Shah 2017). During winter season (December, January and February), Himalayan region experiences copious amounts of rain and snow, and 'Western Disturbances' brings in strong winds with rain, which at times can be heavy. Hence, the mountainous areas of northwest and west India are prone to snow avalanches and landslides with spells of cold waves. The winter season supports heavy fog and hailstorm causing serious damage to crop and orchards. The cyclones and storm surges are the major natural disasters during the pre-monsoon or hot weather season (March, April and May) affecting most of the coastal parts and inlands of Gujarat and Rajasthan. The monsoon season (June, July, August and September) brings rain to entire country with floods wherever the monsoon is more active and droughts wherever the monsoon is weak. The excess rains in mountainous region of Himalaya from Jammu and Kashmir to northeastern states trigger landslides, mudslides, rockslides, etc. The monsoonal rains are responsible for landslides in Western Ghats and the hilly areas of Kerala too. Cyclones are, once again, the major disaster during post-monsoon season (October and November). Northeast Monsoon during the season brings considerable rains to most states of peninsular India making them vulnerable to the threat of floods. Approximately 58.6% of Indian landmass is prone to earthquake (moderate to high intensity), 12% of land is prone to floods and river erosion, 68% of cultivable area is vulnerable to drought and hilly areas are at risk from landslides and avalanches. Consequently, 27 out of 36 states and union territories of India are disaster-prone. In terms of numbers of climate- and weather-related disasters recorded in 2020–2021, the USA has the most (71) followed by Indonesia (62), India (46), China (41), Congo (27), the Philippines (25), Pakistan (25), Mexico (23), Columbia (22) and Peru (21) (World Disasters Report 2022). The ten countries in the world with largest number of fatalities due to disasters in 2020–2021 are India (4743), Nigeria (4299), Haiti (2697), the UK (2559), France (1977), Belgium (1499), the USA (1184), Pakistan (1127), China (1126) and Bangladesh (945). The ten countries in the world with most people affected by disasters in 2020–2021 are China (32.8 million), India (23.5 million), the Philippines (18.6 million), South Africa (12 million), Afghanistan (11.1 million), Bangladesh (9.7 million), South Sudan (9.6 million), Somalia (8.8 million), Ethiopia (8 million) and Iraq (7 million). The 76% of reported climate- and weather-related disasters in the 1960s rose to 83% in 2010–2019 and further to 90.3% in 2020–2021 (World Disasters Report 2020). India, once again, recorded a heavy fatality of 1529 lives between June and September of 2023 due to monsoon floods affecting 10.2 million people (EMDAT Report 2023).

The northwest Himalayan Mountain region is prone to many natural hazards due to its unique geological, climate, and tectonic conditions. Substantial topographic scale and elevation are characteristics of the young mountain region. The enormously folded mountain region gives rise to crumpled and fragile rock successions that are vulnerable to weathering and erosion processes because of alternating ecosystem of snowfall and rainfall, to create secondary landforms of broad U-shaped valleys in higher altitudes and V-shaped valleys in lower reaches. The continuous uplift of the peaks due to ongoing plate collision, adds to the elevation to favour the formation of altitudinal glaciers, which in turn accomplishes rapid erosion of the Himalaya, a phenomenon that has been dubbed the 'tectonic aneurysm' (Zeitler et al. 2001). The

**Table 6.1** Top 10 natural disasters worldwide in 2008 by number of deaths

| Type of disaster (month) | Country | Number of deaths |
|---|---|---|
| Cyclone Nargis (May) | Myanmar | 138,366 |
| Earthquake (May) | China PR | 87,476 |
| Flood (June–August) | India | 1,963 |
| Extreme winter conditions (January) | Afghanistan | 1,317 |
| Typhoon Fengshen (Franck) (June) | Philippines | 644 |
| Hurricane Hanna (September) | Haiti | 529 |
| Mass movement wet (September) | China PR | 277 |
| Flood (October) | Yemen | 180 |
| Flood (June) | China PR | 176 |
| Flood (September) | India | 173 |

*Source* UNISDR (United Nations International Strategy for Disaster Reduction) 2008

continuous uplift of Himalaya coupled with climate change and denudation drives the natural hazards, such as earthquakes, mass movements, avalanches, floods, glacial lake outburst floods (GLOFs), cloudburst, desertification, etc. in the fragile mountain region. The sediment, thus, derived from glacial erosion is carried to the terminus in subglacial streams, which then becomes part of the fluvial (meltwater) system and the cycles of swings in sediment are recorded downstream as cyclical depositional packages. The Emergency Events Database (EM-DAT), an international disaster database, was founded in 1988 by the Centre for Research on the Epidemiology of Disasters (CRED) at the Universit´e catholique de Louvain (UCLouvain, Belgium), to include information on impacts of disasters that have happened since 1900. An analysis of data available on EM-DAT shows that Himalayan regional territories, Ladakh, Jammu and Kashmir, Himachal Pradesh, Uttarakhand (all NW Himalaya) and Eastern Himalaya account for 18% of the geographical area of India, but hotspot for 35% of major hazards over last 110 years (Singh and Pandey 2024). The high Himalayan region is vulnerable to snowstorm and rainstorm induced hazards in the form of avalanches, cloud bursts, flash floods, debris flow landslides, involving repeated heavy loss of life and property.

There are on an average 76 hazard events in the HKH every year, with the highest in China (25) and India (18) (Wang et al. 2019 and reference herein) (Table 6.1). The tectonically active HKH is susceptible to high-intensity earthquakes such as the 1934 Nepal–Bihar earthquake in Nepal (Mw = 8.0), the 2005 Kashmir earthquake in Pakistan (Mw = 7.6), the 2008 Wenchuan earthquake in China (Mw = 7.9) and the 2015 Gorkha earthquake in Nepal (Mw = 7.8) that triggered vegetation degradation, landslides, rockfalls and soil erosion (Ministry of Science, Technology and Environment 2015). Northwest Himalaya is an abode of ~51% of total landslides from 1800 to 2011 in India (Kumar et al. 2021c) and, therefore, hotspot for detailed understanding of the disasters.

## 6.1 Lake Outburst Flood Hazards

Lake outburst floods (LOFs), in the background of change in climate, is both a hazard and major agent of geomorphic change in the Himalaya. The lakes that have been formed behind a landslide dam or in association with a glacier (dammed by a frontal moraine or glacial ice) give rise to LOFs. The mass movements or avalanches into the lake, seismic activity, piping within the dam, overtopping of the dam or degradation of blocking ice (Westoby et al. 2014) can be the cause for catastrophe due to LOFs. The floods, thus generated, are short-lived discharges but with several orders of magnitude higher than the background discharges in the receiving rivers (Cook et al. 2018 and references herein).

### *6.1.1 Glacial Lake Outburst Flood (GLOF) Hazards*

It is widely accepted that planet earth is reeling under the effect of global warming which in turn is responsible for intensified monsoon circulations to cause extreme weather events with changes in the frequency and magnitude. Extreme precipitation (snow or rain) in upstream of Himalayan region often trigger natural hazards, including flash floods, landslides, debris flow and river damming. Large areas in Hindu Kush Himalayan (HKH) region are vulnerable to inundation by flood waters during rainy season every year (Tsering et al. 2021). The upstream of Himalayan rivers in HKH region is highly susceptible to mass failures, due to steep elevation in contrast to short distance, generating landslide and avalanches causing blockage of streams/rivers and sudden outbursts with a slurry of sediments and debris (Gao et al. 2021; Byers et al. 2019).

Glacier lake outburst floods (GLOF) are one of the most serious and unanticipated glacier hazards with potential to release huge amount of water and debris in a short span of time resulting in heavy loss of lives and property, particularly in the downstream. Climate warming has resulted in the retreat and thinning of glaciers in high-altitude of Himalaya leading to the formation of various types of lakes, proglacial, supraglacial, moraine-dammed, etc., of different sizes. Shrestha et al. (2023) delineate occurrence of 29% of GLOFs in the Karakoram, 19% in the Eastern Himalaya and 28% in the Western Tien Shan based on RGI (Randolph Glacier Inventory), whereas delineation is considerably different on the basis of HiMAP (Hindu Kush Himalayan Monitoring and Assessment Programme) where 29% occurred in the Karakoram, 17% in the Northern/Western Tien Shan, 15% in the Central Himalaya and 11% in the Eastern Himalaya. The meltwater ponded for months to decades behind glacier dams or in lakes occupying tributary valleys or behind moraine dams, can cause large outburst floods (GLOFs) that may have peak discharges of the magnitude greater than those associated with weather-related floods, and occurrence of a unique hazard (Table 6.2).

**Table 6.2** Types of glacier, glacial and related hazards (after Richardson and Reynolds 2000)

| Category | Hazard event | Description | Time scale |
|---|---|---|---|
| Glacier hazards | Avalanche | Slide or fall of large mass snow, ice and/or rock | Minutes |
| | Glacier outburst | Catastrophic discharge of water under pressure from a glacier | Hours |
| | Jökulhlaup | Glacier outburst associated with subglacial volcanic activity | Hours-days |
| | Glacier surge | Rapid increase in rate of glacier flow | Months-years |
| | Glacier fluctuations | Variations in ice front positions due to climatic change, etc. | Years-decades |
| Glacial hazards (*as above, plus:*) | Glacial lake | Catastrophic outburst from a proglacial lake, typically | Hours |
| | Outburst floods | Moraine dammed | |
| | Débâcle | Outburst from a proglacial lake | |
| | Aluvión | Catastrophic food of liquid mud, irrespective of its cause, generally transporting large boulders | |
| Related hazards | Lahars | Catastrophic debris #ow associated with volcanic activity and snow fields | Hours |
| | Water resource problems | Water supply shortages, particularly during low #ow conditions, associated with wasting glaciers and climate change, etc | Decades |

Overall increase of 5.9% in number of lakes and 6.8 ± 0.1% in area is reported in HMA (High Mountain Asia) between 1990 and 2015 (Zheng et al. 2021b) while Wang et al. (2020) reported increase by $n = 2916$ and 273.65 km$^2$ between 1990 and 2018 and Chen et al. (2021) reported increase by $n = 3342$ and 220.64 km$^2$ between 2009 and 2017. The increase in number and area is driven by the expansion of proglacial moraine-dammed glacial lakes (Zheng et al. 2021b). Spatio-temporal analyses of selected moraine-dammed lakes in Satlej basin in western Himalaya from 1990 to 2018 show increase in lake area from 0.11 to 0.26 km$^2$ over past 28 years (Rawat et al. 2023a). The sudden advance of Chong Khumdan surging glacier into

Shyok River in Karakoram formed series of large ice-dammed (jökulhlaups) and outburst lakes between 1926 and 1932.

In Indian Himalayan region (IHR) covering Indus, Gang and Brahmaputra River basins, total of 28,043 glacial lakes have been mapped and classified into four major categories namely moraine-dammed, ice-dammed, glacier erosion and other glacial lakes, based on the process of lake formation, location and type of damming material (NRSC 2024). Although glacial lakes are important water resources for downstream communities (Rawat et al. 2023b) yet increase in the number and areal extent of glacial lakes brings both large opportunities and risks to the lives and livelihoods in the form of catastrophic GLOF events (Dubey and Goyal 2020). Pamir and Karakoram regions mostly experience ice-dam lake failures (Bhambri et al. 2019), whereas moraine-dam lake failures are common in the IHR (Nie et al. 2018). GLOFs in the IHR are one of the most serious natural hazards (Chen et al. 2021) due to the steep slopes and narrow flow channels in the terrain (Sattar et al. 2019) and have caused maximum number of casualties in the Central Himalayas despite fewer floods.

Localized and sudden downpour phenomenon of rainfall, called cloudburst, results into devastating flash floods. The cloudburst, a natural hazard, in the Himalayan terrain (Table 6.3) is invariably observed in the form of flash floods and sudden debris flow following the heavy precipitation in very less time interval over a very small area. The events of cloudburst in mountainous regions are associated with cumulonimbus or thunder clouds that are a result of interaction of low-level westward moving monsoonal systems and eastward moving mid-tropospheric westerly trough. The cloudburst, rapid melting of snow or ice and heavy rainfall in September 2023 in Sikkim led to massive flooding downstream (Dahal et al. 2024) causing catastrophic event. Calving of Lhonak lake glacier snout resulted in the collapse of section of left lateral moraine that facilitated rapid outflow of water causing erosion and widening of lake outlet, partial blockage of Teesta River due to landslide at Lachen, and straightness of the Teesta River channel from the confluence of Lhonak Chu and Teesta River up to Chungthang together magnified the destructive force of breached water from the lake. Recent catastrophes in Melamchi River (Koshi River basin, Nepal) in 2021 is a case study of several (geomorphologic, hydroclimatic and physical) compounding factors resulting in the cascading effect (Adhikari et al. 2023).

A few events of GLOFs are reported from the Trans-Himalayan region of Ladakh during recent decades, from Nyoma in the Changthang area in eastern Ladakh in 1971, from Domkhar in 2003 and recently the Gya flood in 2014 (Schmidt et al. 2020 and references herein). The spatio-temporal analyses show that Gya proglacial lake expanded continuously from $0.03 \pm 0.002$ km$^2$ in July 1969 to $0.12 \pm 0.003$ km$^2$ in August 2019 with a rapid increase in the lake area to double from $0.06 \pm 0.007$ km$^2$ to $0.12 \pm 0.003$ km$^2$ since 2005. A drastic short-term lake-level rise occurred with an area increase of about 0.02 km$^2$ to $0.11 \pm 0.002$km$^2$ from July 2014 to 6 August 2014, the day of the flood. The analyses also suggest that the number and size of glacial lakes have increased since 1969, an observed trend in the entire Himalayan region (Bolch et al. 2019). The relative increase in the total glacial lake area, which amounts to 35% in the trans-Himalayan region of Ladakh between 1993 and 2018,

**Table 6.3** Details of cloudburst events reported over the southern rim of the Himalaya (after Dimri et al. 2017)

| S. No. | Date and time | Location (Lat./Lon./Alt.) | Precipitation reported (mm/h) | Damage (Death, loss of property, landslides, flash-flood etc.) |
|---|---|---|---|---|
| 1 | 20 Jul 1970 | Ganai-Almora, Uttarakhand, 29.88 °N, 79.35 °E, 1900 m | | 200 humans, 33 houses |
| 2 | 17 Jun 1979 | Saikot-Chamoli, Uttarakhand, 30.4 °N, 79.3 °E, 2200 m | | 3 humans, 70 animals, 50 houses |
| 3 | 17 Aug 1979 | Kuntha-Rudraprayag, Uttarakhand, 30.4 °N, 79.05 °E, 1300 m | | 39 humans, 39 animals, 20 houses |
| 4 | 17 Aug 1979 | Sirwari-Rudraprayag, Uttarakhand 29.9 °N, 79.09 °E, 1700 m | | 13 humans,150 animals, 34 houses |
| 5 | 31 Jul 1982 | Mandakhal Chennil-Pauri Garhwal, Uttarakhand, 29.8 °N, 78.7 °E, 1700 m | | 3 humans, 80 animals, 8 houses |
| 6 | 22 Jul 1983 | Karmi-Kapkote**,** Bageshwar, Uttarakhand 30.1 °N, 79.8 °E, 1981m | | 150 humans, 20 animals, 6 houses |
| 7 | 09 Jul 1990 | Nilkanth-Puri Garhwal, Uttarakhand, 30.7 °N, 78.6 °E, 1200 m | | 100 humans, 10 houses |
| 8 | 16 Aug 1991 | Dewar Khadora, Gangolgaon, Chamoli, Uttarakhand, 30.4 °N, 79.2 °E, 1600 m | | 24 humans, 63 animals,38 houses |
| 9 | 2 Sept 1992 | Gadni-Chamoli, Uttarakhand, 1900 m | | 14 humans, 31 houses |
| 10 | 13 Aug 1995 | Bhintai-Pauri Garhwal, | | 13 humans, 6 houses |
| | | Uttarakhand, 30.1 °N, 78.7 °E, 1300 m | | |
| -73152-52133411 | 17 Jul 1996 | Berinaga Pithoragarh, Uttarakhand, 29.7 °N, 80.05 °E, 2100 m | | 18 humans, 85 animals, |
| 12 | 11 to 19 Aug 1998 | Okhimath-Rudraprayag, Uttarakhand, 30.55 °N, 79.21 °E, 1900 m | | 103 humans, 422 animals, 820 houses |

(continued)

**Table 6.3** (continued)

| S. No. | Date and time | Location (Lat./Lon./Alt.) | Precipitation reported (mm/h) | Damage (Death, loss of property, landslides, flash-flood etc.) |
|---|---|---|---|---|
| 13 | 18 Aug 1998 | Malpa-Alaknanda, Uttarakhand 3000 m | | 221 humans, 60 animals, 40 houses |
| 14 | 11 Aug 2001 | Phata-Rudraprayag, Uttarakhand 30.57 °N, 78.9 °E, 1800 m | | 27 humans, 64 animals, 22 houses |
| 15 | 31 Aug 2001 | Gona, Tehri, Uttarrakhand 30.36 °N, 78.68 °E, | 22 | 7 humans, 7 animals, 28 houses |
| 16 | 16 Jul 2003 | Shillagarh, Himachal Pradesh, 31.5 °N, 77 °E | 75–100 | 35 humans |
| 17 | 6 Aug 2010 | Leh, Jammu & Kashmir 34.09 °N, 77.34 °E | 23.3 | 255 humans, 1749 houses |
| 18 | 3 Aug 2012 | Pandrasu ridge,Garhwal, Himalaya Uttarakhand | | 35 humans, 436 livestock lost, 591 houses |
| 19 | 13 Sep 2012 | Ukhimath,Rudraprayag, Uttarakhand 30.5 °N, 79.25 °E | 23.3 | 66 humans |
| 20 | 16-17 Jun 2013 | Badrinath, Gangotri, Ymunotri and Kedarnath, Uttarakhand 30.7°-30.8 °N, 79°-79.1 °E | 25 | 6600 humans, 365 houses, |
| 21 | 16 Jul 2015 | Sonamarg, Srinagar, Jammu & Kashmir, 34.40 °N, 74.71 °E | | 4 humans |
| 22 | 12 Jul 2015 | Pahalgam, Anantnag, Jammu & Kashmir, 34.01 °N, 75.31 °E | | 2 humans |
| -73152-626923 | 17 Jul 2015 | Ganderbal, Jammu & Kashmir 34.21 °N, 74.77 °E | | 3 humans |
| 24 | 24 Jul 2015 | Baltal, Jammu & Kashmir, 34.6 °N, 74.7 °E | | Not reported |
| 25 | 28 Jul 2015 | (a) Choskore village, Kargil, J & K 34.19 °N, 76.35 °E (b) Wakha villages, Kargil, J & K, 34.16 °N, 77.6 °E | | 20 houses |

(continued)

**Table 6.3** (continued)

| S. No. | Date and time | Location (Lat./Lon./Alt.) | Precipitation reported (mm/h) | Damage (Death, loss of property, landslides, flash-flood etc.) |
|---|---|---|---|---|
| 26 | 25 Jul 2015 | Jibhi village, Banjar Division Kullu, Tandi Nullah 31.63 °N, 77.34 °E | | Not Reported |
| 27 | 22 Jul, 28 Jul, 9 Aug 2015 | Biama, Leh, Jammu & Kashmir 34.15 °N, 77.57 °E | | Office-cum residence of Water and Power Consultancy Services (WAPCOS) damage |
| 28 | 4 Aug 2015 | Upper Ganglass, Saboo, etc. Leh, Ladakh, J & K 34.31 °N, 77.53 °E | | Not Reported |
| 29 | 8 Aug 2015 | Dharampur, Mandi, Himachal Pradesh 31.7 °N, 76.7 °E | | 5 humans |
| 30 | 28 May 2016 | (a) Kemra, Kothiara, Tehri Uttarakhand, 30.57 °N, 78.68 °E, (b) Siliara, Kothiara,Tehri Uttarakhand, 30.46 °N, 78.62 °E, | | (a) Kemra, Kothiara, Tehri Uttarakhand, 30.57 °N, 78.68 °E, (b) Siliara, Kothiara,Tehri Uttarakhand, 30.46 °N, 78.62 °E, |

is lower than the observed trend for the eastern Himalaya and Spiti Lahaul (Nie et al. 2017, Sharma et al. 2018b) with only 6% larger than 100,000 $m^2$. Two largest lakes in Ladakh region are identified in 2018, one of them is in eastern Ladakh, 24 km to the south of Pangong Lake, and the second one is in the western part, close to the confluence of Shyok and Indus River. The studies further indicate that 22 glacial lakes between 1969 and 2018, four between 1969 and 1993, nine between 1993 and 2002 and nine from 2002 to 2018 disappeared/reduced considerably in size. Most of these lakes were in the sparsely populated Changthang region in eastern Ladakh. The heterogeneous pattern of changes in lakes of Ladakh region and recurrent fluctuations is similar to observations made in adjacent Karakoram (Ashraf et al. 2017), Koshi basin (Shrestha et al. 2017) and along Himalayan arc (Nie et al. 2018). Geological Survey of India identified 13 vulnerable lakes out of 486 in the Uttarakhand Himalaya (Rana et al. 2021) with eight glacial lakes identified as critical and having outburst potential.

The Kedarnath disaster, primarily attributed to the catastrophic breach of Chorabari lake, is considered to be the role of snowmelt and its contribution to runoff as evident from MODIS (Moderate-Resolution Imaging Spectroradiometer) derived snow cover mapping in the Kedarnath catchment showing that streams in the region were flowing at unusually high levels for several weeks before the onset of the heavy rains in mid-June of 2013 (Allen et al. 2015). The unusually high runoff contribution was from the melting of the seasonal snowpack.

The Kedarnath GLOF event was induced by heavy rainfall (Rafiq and Mishra 2018) and the Chamoli disaster was a result of avalanche (Shugar et al. 2021) in Indian Himalaya suggesting that numerous factors, such as dam stability, overtopping, an extreme climate event such as heavy rainfall, seismic activity, avalanches and/or combination of them, can trigger a GLOF. The increase in maximum temperature and decrease in winter precipitation in Himalaya are attributed to atmospheric warming and loss of glacier ice that contribute towards change in size and shape of glacial lakes in the higher elevations. Other parameters, like dam characteristics, the parent glacier, topographic potential, tectonic movement, heavy precipitation and population hotspot in the downstream add to the GLOF hazards (Islam and Patel 2022). GLOF event in Himalaya are closely associated with rapidly growing moraine-dammed glacial lakes which are of usually elongated shapes compared to newly formed circular lakes (Peng et al. 2023; Wang et al. 2024). Moraine-dammed lakes remain dormant for long period, spanning several decades, to become inhabited gradually and expose life and property to potential GLOF disaster whereas ice-dammed lakes are geographically located distantly to cause major disasters though 70% of all reported GLOF are outburst of ice-dammed lakes (Emmer 2024).

The increment in glacier associated hazards is attributed to early onset of snow melting period driven by noticeable increase in temperature trends supporting vulnerability of the Himalaya towards drier climatic extreme (Gujree et al. 2022; Mushtaq et al. 2022).

### *6.1.2 Landslide Lake Outburst Flood (LLOF) Hazards*

The complex interaction between climate and geomorphic processes in mountain environments, such as Himalaya, give rise to landslides and other mass movements (particularly rockfalls) which, at times, temporarily block or dam a river to form lakes which can breach at timescales varying from days to years after their formation and produce devastating landslide lake outburst floods (LLOFs) (Ruiz-Villanueva et al. 2017 and references herein). The focus of study, so far, has been on the impacts of LLOFs on efficiency in transporting the sediments, channel erosion, bank undercutting and parallel retreat (Cook et al. 2018; Turzewski et al. 2019) based on retrospective field reconnaissance or post-event satellite image interpretations. However, there is lack in understanding the geomorphic processes associated with LLOFs causing the development of devastating hazards and impact of LLOFs in the modification of landscapes. The rock avalanches or slides, and debris flows in mountainous regions can block narrow and steep walled river valleys and impound the upstream with highly variable timings of sudden outburst or longevity of landslide dams from few minutes to hundreds of years (Shrestha and Nakagawa 2016). Climatically driven triggers are linked to patterns of precipitation, temperature and insolation, and non-climatically driven triggers, which include mechanical failures of the dam as well as surrounding hill slopes, are capable of causing a sudden lake outburst on their own or in combination or even facilitate one another (cascading triggering mechanisms) (Fischer 2022 and references herein). The rate of ice-mass loss nearly doubled in glaciers which had a terminus in contact with glacial lakes compared to those terminating on land, facilitating the growth in number, area and depth of glacial lakes (King et al. 2019). The total glacial lake area expanded by 14% between 1990 and 2015 in the Himalayas (Nie et al. 2017) and by 15% between 1990 and 2018 across whole High Mountain Asia (Wang et al. 2020).

The Yigong outburst of rockslide dammed lake in 2000 is a prominent example of LLOF in the Greater Himalayas, affecting major transportation infrastructure and life in China and northern India (Delaney and Evans 2015). The Melamchi Khola in Nepal, presumably landslide-dammed lake outburst, is a more recent example causing damage to numerous bridges impacting the drinking water supply heavily (Maharjan et al. 2021). Two successive landslides blocked the Jinsha River near Baige village (eastern Tibet) twice in October and November 2018 and two outburst floods generated after overtopping of the dam (Liu et al. 2021a) causing bank undercutting and parallel retreat which led to the destabilization of at least nine hillslopes by way of removing their buttresses (Yang et al. 2021).

The catastrophic floods in the Spiti River and upper Sutlej basin, Himachal Pradesh in July 2000 and June 2005 are the cases of LLOFs in Himalaya. The breach in Parchu Lake (also called Parechu or Paree Chu or Parachu), Tibetan Plateau approximately after 12 months of damming of lake due to landslide activity is considered a major cause for 2005 LLOF (Ruiz-Villanueva et al. 2017). Contributions from significant snowmelt, resulting primarily from an extremely large precipitation event in late January to early February, also served as crucial hydrometeorological factor leading to the dam breach.

The magnitude and recurrence intervals of LOFs impacts the channel and adjacent hillslopes of the mountainous river valleys by way of fluvial incision that sets fresh base level for the entire landscape. The non-climatic (e.g. earthquake) and climatic (e.g. air temperature, variability of the Equilibrium Line Altitude (ELA) and, to a degree, the glacial recharge) factors are considered as the noticeable drivers in affecting the size and distribution of glacial lakes and consequently the erosion rates in a LOF susceptible regions, instead of the strength of monsoon or measures of precipitation (Schwanghart et al. 2016, Bricker et al. 2017). The relationship between fluvial erosion and precipitation becomes complicated and nonlinear even in the cases where LLOF is linked to precipitation.

## 6.2 Landslide Hazards

Landslides are one of the most devastating and recurring natural disasters and have affected several mountain regions across the globe. The Indian Himalayan region is no exception to landslides incidence affecting key economic sectors such as transportation and agriculture and often leading to loss of lives. According to the global landslide disaster database, a total of 5318 non-seismic landslides occurred from 2004 to 2017, of which 3285 landslides were triggered by rainfall (Froude and Petley 2018) while 580 landslides occurred in the Indian Himalayas during the same period, of which 477 were triggered by rainfall, thereby contributing 14.52% of the global landslides (Dikshit et al. 2020). The major landslide event in Mandakini valley, Uttarakhand in August 1998 comprised 466 landslides in total and was triggered by heavy rainfall followed by another landslide event due to cloudburst in July 2001 and major landslide event comprising 473 landslides in September 2012 due to heavy rainfall (Dikshit et al. 2020 and reference herein). Heavy rainfall, 350 mm in five days, in June 2013 was a cause for landslides in Chamoli and Rudraprayag districts of Uttarakhand. Himachal Pradesh has a history of recurring major landslides since 1977, particularly the Katropi area (Roy et al. 2018, Pradhan et al. 2019), whereas the Pawari landslide zone is active and has expanded by 7% during 2005–2014 (Kumar et al. 2018). Analysis revealed episodes of cascading motions triggered by extreme rain events resulting an overall increase in landslide velocity from $\sim$ -22 mm/year during 2004–2010 to $\sim -325$ mm/year during 2022–2023 that finally culminated into series of catastrophic land deformation events at Joshimath (Chamoli district, Uttarakhand) between December 2022 and January 2023 (Sreejith et al. 2024). Major cloudbursts such as the Leh Nalla cloudburst (2005, 2006), Leh cloudburst (2010), and Baggar cloudburst (2011) are the noticeable events of landslides and flash floods in Ladakh region (Banerjee and Dimri 2019). Though the Ladakh region is shadowed from Indian Summer Monsoon yet is highly sensitive and susceptible to mass movement events. The unconsolidated clasts created by the mechanical weathering of the Ladakh batholith and Zanskar metasediments on the hill slopes of Zanskar and Ladakh ranges is transported downslope by meltwater from the glaciers, by the glaciers themselves or by the process of creep giving rise to landslide and/or rockslide. In recent past, four large disasters of cloudburst, landslides and rockfall occurred in August 2010, July 2012, May 2013 and August 2013 in Ladakh region. Geological archives supporting palaeo-lake outburst are preserved at Spituk (Leh) (Mujtaba et al. 2018) confirming that seismic activity, climate oscillations and the critical hydrostatic conditions resulting from geomorphic in-equilibria could have caused the changes in hydrostatic pressure. Darjeeling region in northeast Himalaya is seriously affected by landslides. Since the first recorded occurrence of landslides in 1899, reports of recurrences are in 1950, 1968, 1991, 1993, 2004, 2005, 2006, 2009 and 2015 (Mondal and Mandal 2019, Dikshit and Satyam 2019). Similarly, south Sikkim area is prone to numerous landslides (Bera et al. 2019). The National Highways 40, 44 in Meghalaya, and 44-A in Mizoram have faced several landslides in 2011, 2013 and 2017 (Sardana et al. 2019). Two flooding events (bracketed between 8.5 ± 0.4 ka and 7.5 ± 0.4 ka), based on OSL dating, are recognized in Shyok-Nubra valley of Ladakh (Hakhoo et al. 2019). Studies of landslide at Nainital (Uttarakhand) infer that the weak rock type of Infra-Krol Formation, characterized by extensive presence of foliation planes and steep valley dipping joints, is mainly responsible for extensive weathering effects along these geological discontinuities which in the high annual precipitation zone with steep stream bed gradients and steep side slopes contribute to instability and mass wasting.

Besides GLOF, NW Himalaya is highly susceptible to floods as a consequence of alterations in river patterns due to geological instability, melting of glaciers, extreme monsoonal rainfall and rapid urbanization (Sultan et al. 2022). The Kashmir valley, in NW Himalaya with long historical records of high-impact floods (Ballesteros-Cánovas et al. 2020), is highly vulnerable to floods having impact on socio-economy and environment (Malik 2022). The geomorphic configuration of the Jhelum River basin exacerbated by anthropogenic factors such as urbanization and loss of wetlands (Romshoo et al. 2018), and watershed characteristics (terrain and land cover) influenced the floods in 2014, worst in last hundred years (Malik et al. 2024). OSL dated flood events at around 6.5, 4.5, 2.6, 1.4, 0.8 and 0.4 ka in upper Ganga catchment correspond to periods of moderate ISM suggesting a coupling between warm-humid monsoon and relatively dry Westerlies (Sharma et al. 2022b). The flash floods of 1902, 1945, 1988, 1993, 1995, 2000 and 2003 in the Kullu valley (Himachal Pradesh) are related with the extreme rainfall and occasional cloudburst events during the monsoon period from mid-July to mid-September. Deforestation, slope cutting and heavy rainfall are key driving factors causing soil erosion, landslides and rockfall in the Alaknanda River basin. Besides, overgrazing and tourism are identified as other factors for rockfall hazards and soil erosion (Pandey and Prasad 2018). The slope cutting for the construction of roads and hotels, buildings and other tourism-related facilities, directly affects the runoff and soil erosion. Nearly 28% of Alaknanda River basin is with steep slope supporting shallow soil and less forest cover facilitating soil erosion, landslides and hydrological hazards. Also, maximum frequency of landslides in Alaknanda River basin occurs along Madhyamehashwar, Godwanala and Rawanganga faults.

## 6.3 Earthquake Induced Hazards

The average rate of convergence between the Indian Plate south of the Himalayan foothills and southern Tibet at Main Himalayan Thrust, a 2000 km long and approximately 100 km wide décollement separating the Indian and Asian plates, is 17 mm $a^{-1}$ accumulating seismic moment at a rate of c. 1020 $Nma^{-1}$ (Newton metres per year) equivalent to that released by a Mw = 7.3 earthquake (Bilham 2019), mostly at shallow depth (< 30 km). Archival documents (newspapers and travelogues) and excavations of earthquake-related features (surface liquefaction or the offset of surface faults) give an account of the historical earthquakes in the Himalayan region. Major earthquakes in sixth and fifth century BCE, during the life of Buddha, at Lumbini (Nepal) and rupture of Himalayan Frontal Thrust at Ramnagar are chronometrically established (Malik et al. 2017). Critical examination of the pre- and post-damaged photographs, published in archival documents, provides a historical evidence of the incremental damage to the temples in Kashmir and Kangra due to earthquakes in the region (Szeliga and Bilham 2017). Medieval earthquakes at 1100BCE and 1255BCE in India, Nepal and Bhutan are recorded along Himalayan Frontal Thrust as palaeo-seismological events of surface faulting (Wesnousky et al. 2017).

The Indian subcontinent has experienced major earthquakes of varied magnitude because of the tectonics in Himalayan region causing serious hazards. One of the major earthquakes in the last two centuries is the Assam earthquake (magnitude = Mw 8.1) of 1897 triggering around 27 landslides on the Shillong plateau and Great Assam earthquake (magnitude = Mw 8.6) in 1950 triggering around 1,00,000 landslides (Singh and Som 2016). Two major earthquakes in Uttarakhand occurred in Uttarkashi (1991) and Chamoli (1999) of magnitudes 6.6 and 6.8, respectively leading to 23 landslides in 1991 and 56 landslides in 1999 (Sangeeta and Maheshwari 2019). About 2250 landslides were triggered in India and Pakistan collectively because of the Kashmir earthquake (magnitude = Mw 7.6) that struck Muzaffarabad (Pakistan) in 2005 (Joshi and Subramanian 2023 and references herein). The Sikkim earthquake (magnitude = Mw 6.9) in 2011 resulted in reactivating earlier landslides and 1196 landslides were triggered in Sikkim and North Bengal (Singh and Singh 2016). The recent Gorkha earthquake (magnitude = Mw 7.8) in 2015 affected 2million people in Nepal, China and India (Hall et al. 2015). The movement along the major thrusts—Main Central Thrust (MCT), Munsiari Thrust (MT) and Berinag Thrust (BT), and subsidiary faults—occurring in the Eastern Ramganga River basin, Kumaun Himalaya, have been responsible for shearing, crushing and pulverization of rock, which are easily susceptible to rains and imprudent activities of human (Joshi et al. 2021) (Table 6.4). The basin is an earthquake-prone area in the vicinity of thrusts where small to medium magnitude earthquake induced landslides, particularly during the earthquakes of 1879, 1916, 1958 and 1980 have ravaged the slopes.

Moderate to great earthquakes in the Himalayas—Assam (1897), Kangra (1905), Bihar-Nepal (1934), Shillong (1950), Bihar-Nepal (1988), Uttarkashi (1991), Chamoli (1999), Kashmir (2005), Central Nepal (2015), etc.—are responsible for the slope failures triggering disaster and casualties (Prasath et al. 2019 and references herein). The movements along major tectonic boundaries in Himalaya (South Tibetan Detachment System—STDS, Main Central Thrust—MCT, Main Boundary Thrust—MBT, Himalayan Frontal Thrust—HFT) and their splays are responsible for generating earthquakes and associated landslide hazards (Singh et al. 2022a).

**Table 6.4** Damage history of temples in Chamba province, Himachal Pradesh (after Joshi and Thakur 2016)

| S. No. | Name of the temple | Location | Age of construction | Deformation feature | Deformation direction | Restoration age of temple | Earthquake |
|---|---|---|---|---|---|---|---|
| 1 | Lakshmi Narayan | Chamba | Tenth century | Tilting and cracks of temple pillars, displacement of the supporting pillars, upwarping of the floor stone | North–south | 1582 and later | 1555 Kashmir |
| 2 | Shiv temple | Chotra, Chamba | Not known | Tilting of the entire temple structure | S20°E | Not restored | 1555 Kashmir |
| 3 | Shiv temple | Near Sanatan Dhram Sanskrit Mahavidal ay, Chamba | Not known | Tilting of the entire temple structure | N20°W | Not restored | 1555 Kashmir |
| 4 | Shiv temple | Kariya, Chamba | Not known | Tilting of the temple and structure | South | Partially recently restored | 1555 Kashmir |
| 5 | Chamunda Devi temple | Chamba | A.D. 1762 | No deformation | - | - | - |
| 6 | Chitrari temple | Chitrari | A.D. 680 | Tilting of pillar and main door | N80°E | Partially recently renovated | 1905 Kangra |
| 7 | Bharmour group complex | Bharmou r | A.D. 700-tenth century | Tilting and cracks of temple pillars, displacement of the roof stone | N80°W | Partially recently restored | 1905 Kangra |

## 6.4 Snow Avalanche Hazards

High Mountain Asia region is at the risk of snow avalanches triggered by strong ground motions and meteorological factors. Collapsing or avalanching glaciers and glacier detachments or calving are the two types of avalanches commonly in High Mountain Asia region (Kääb et al. 2021). The collapsing or avalanching glaciers are the falling of ice masses or seracs over glacier ice flows on a steep slope or along a ridge because of a tensile failure of the ice induced by the combination of internal flow (creep) and sliding of the glacier facilitated by contribution of melt water as well as extra load on top of the ice. The glacier detachments or calving results when a large piece of the glacier tongue detaches to form an avalanche. The glacier instability gives rise to ice-snow avalanches, ice-rock avalanches, glacier detachment and glacier surges which are driven by the velocity of glacier (Bhambri et al. 2017). The avalanche activity is modified with change in climatic conditions and land cover, such as afforestation and deforestation (García-Hernández et al. 2017).

The avalanche formation zones in Indian Himalaya are at high altitudes (> 4000 m amsl) above the tree line and on critical slopes between 30° to 45° whereas studies conclude that release areas lie between 25° and 40° (Kumar et al. 2017; Singh et al. 2018). Central Himalayan region experiences avalanche events during the pre-monsoon dry periods due to higher temperature variability favouring wet avalanches and during post-monsoon due to weakened layers from the warm monsoon season. The heavy winter precipitation due to Western Disturbances during the winter season results excessive snow loading on snow covered terrain in western part of the HKH leading to the occurrences of most avalanches (Prakash et al. 2014). Rise in winter temperature during winter and early spring in western Himalaya favours the wetting of snow and increase in the occurrences of wet snow avalanches contradicting that warming leads to less snow, and hence lower avalanche activity (Ballesteros-Cánovas et al. 2018). The increased avalanching in the western Indian Himalaya is likely to result from the preservation of a sufficiently thick snow cover in high-elevation avalanche release areas combined with more frequent crossing of the melting point of snow, which, in turn, results in more frequent, mostly wet, snow avalanches (Ballesteros-Cánovasa, et al. 2018).

A few important avalanche accidents in India are, 258 killed in Lahaul (Himachal Pradesh) in 1979, 207 persons killed in Zojila Pass (Jammu and Kashmir) tragedy in 1988, 40 army persons killed in Kargil (Jammu & Kashmir state, now Union Territory of Ladakh) in 1994 (Fig. 6.1), 70 persons died on national highway (Jammu and Kashmir) in 1995 and 20 persons feared killed in Sikkim in 1998. The reconstruction of snow avalanche in Hindu Kush-Karakoram-Himalaya region, based on tree-ring analyses, reveals that no avalanche activity took place between the 1940s and 1960s, whereas very high activity took place between 1970 and 1977 and again between 1989 and 2003 with occurrence rates > $0.875y^{-1}$ (Ballesteros-Cánovasa et al. 2018) including four extensive snow avalanche events at Lahaul (Himachal Pradesh) that impacted trees of more than one transect in 1985, 1997, 2000 and 2004.

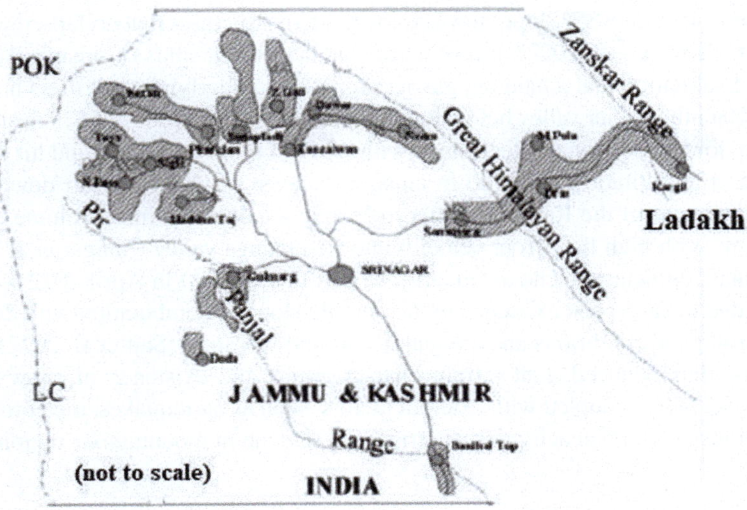


**Fig. 6.1** An overview of avalanche prone areas of part of western Himalaya (after Singh et al. 2004)

The avalanches that occurred on the same day event in Himachal Pradesh, Kashmir and the Ladakh (western Himalaya, India) are found to be a direct contribution of seismic tremors to avalanche release in the area and could not be explained by nival–meteorological factors. The local earthquakes of magnitude $1.7 \leq Mw \leq 4.4$ in Nubra-Shyok region (Ladakh) during 2010–2012 triggered 81 seismogenic snow avalanches up to 92kms from the epicentre of earthquakes (Parshad et al. 2019). Two notable avalanche events in HMA (High Mountain Asia) were triggered by seismic movement, namely the avalanches during the 2015 Gorkha earthquake (Fujita et al. 2017) and the Lenin Peak avalanche in 1990.

Increase in the frequency of ice/snow avalanches is attributed to glacier instability, a factor of disturbance in water-heat balance of glacier due to the fluctuating precipitation and continuous warming of the Himalaya (Li et al. 2024). The continuous warming of Himalayas, in future, will shrink the glaciers, induce meltwater from glaciers, change distribution of water and heat from surface to bed of the glacier and reduce the resistance by bedrock to predispose a larger volume of ice-snow avalanches (IPCC 2019). Increase in the number of ice-snow avalanches from 10 in 1988 to 27 in 2020 with reduction in average deposit area by approximately 70%, from $3.48 \times 10^5 m^2$ in 1988 to $1.2 \times 10^5 m^2$ in 2020 in Annapurna II glacier and Nielamu region (Central Himalaya) from 1980 to 2021 is attributed to increase in annual average temperature at 0.8 °C ± 04 °C (Li et al. 2024). The increase in ice-avalanches is closely related to the increase in glacier velocity.

The process of crack development in, as early as, 2015 culminated in a deadly rock and ice-avalanche causing devastation in Rishiganga–Dhauliganga valley of Chamoli district, NW Himalaya in February 2021. The mixture of rock and ice travelled with extremely high velocity flow front (initiated with ~ 55 m/s speed) from the source (average 35° slope) towards the modern structures that collapsed with big impacts (Tiwari et al. 2022). It is observed on the basis of study of seismic data with ground validation that a hanging glacier along with underlying rock mass broke off in the Raunthi glacier valley below Raunthi peak from an altitude of 5525 m amsl and moved downward along a steep slope with relief of more than 2 km and hit Raunthi stream at an altitude of ~ 3500 m amsl, a distance of about 1.6 km downstream from the snout of the Raunthi glacier, releasing a calculated mass volume of ~ 27 $\times 10^6 m^3$ with high flow front velocity into Rishiganga valley (Shugar et al. 2021). The catastrophic avalanche at Gayari (NE part of Pakistan) in April 2012 was triggered due to snowpack existence on favourable slope in combination with different meteorological conditions and anomalous ground vibration (Saif et al. 2022).

It is, thus deduced, that terrain characteristics and existence of snow depths over weak layers coupled with external factors, such as earthquakes, meteorological conditions and human activities can trigger avalanche in mountainous regions.

## 6.5 Forest Fire Hazards

Western Himalaya envelops Jammu and Kashmir, Himachal Pradesh, Uttarakhand and hill parts of Punjab and Haryana with forest types from alpine forests to semi-evergreen, deciduous, subtropical broad-leaved hill forests, subtropical pine forests and subtropical montane temperate forests. Major fires in Uttarakhand and Himachal Pradesh have been reported in 1911, 1921, 1930, 1931, 1939, 1945, 1953, 1954, 1957, 1958, 1959, 1961, 1964, 1966, 1968, 1970, 1972, 1995, 1999, 2010, 2012 and 2016 (Dobriyal and Bijalwan 2017). Abundant availability of dry Chirpine needles and dry leaf litter of other broad-leaved trees on the forest floor is the major reason for forest fire in the hilly region. Using USGS and NASA FIRMS portals, the forest fire areas of 2022 were mapped, revealing highly sensitive zones in river valleys and mid-altitudes below 1800 m amsl, covering 13% of the total forest area hosting, primarily, monsoon deciduous and flammable pine forests (Sati 2024). The pre-monsoon period (March–May) is considered to be high-temperature tenure in Nepal that results in drought and forest fires (Matin et al. 2017). Above two-thirds (86%) of the forest fires occur during the pre-monsoon season (March to May) in Nepal, of which broad-leaved closed forests have 67% of the total forest fire in Terai Arc Landscape (TAL) areas followed by broad-leaved open forest (12%) (Parajuli et al. 2020). Highest forest fires were in broad-leaved closed forest in April 2016 followed by March 2009. The northwestern Himalaya, in recent decade, witnessed experienced devastating forest fires, particularly during the hot and dry pre-monsoon (March to June) period (Dobriyal and Bijalwan 2017) increasing the burn area between 2001 and 2019 in Uttarakhand and Himachal Pradesh at a rate of 72.94 $km^2\ y^{-1}$ (Bar et al. 2021). The incidents of forest fires are confined within ~ 2000 m amsl of elevation where near surface temperature is comparatively higher (Bar et al. 2023). Although, there is inclination of fire incidences with the increase in land surface temperature and it declines with the increase in rainfall, yet other variables also play a significant role in producing the forest fires, such as soil moisture content, anthropogenic interferences, topographical factors, phenology, etc. (Kumar and Kumar 2022).

FSI (2019) identifies nearly 4% of the forest cover extremely fire-prone, 6% highly fire-prone and 36% prone to frequent forest fires in India. Approximately 65% of the deciduous forests of India are prone to forest fire (Ashutosh et al. 2019). Out of the 192 forest fires during May 2019 in India, the highest (91) were reported from Uttarakhand, were attributed to sudden rise in temperature and a prevailing rainfall deficit during the preceding week (Mina et al. 2023). The availability of dry vegetation biomass favours peak in forest fires in Uttarakhand in late April and late May/early June from 922 forest fires in 2002 to 41,600 in 2019 (Gupta et al. 2018). Highest number of forest fires (8555) were reported from Tehri, Garhwal district whereas fewest (39) were reported from Rudraprayag district of Uttarakhand in 2019. High atmospheric temperatures and dryness (low humidity) favour the production of dry vegetation biomass susceptible to catch fire in the forests (Savita et al. 2018) in addition to high resin content in tree trunks in coniferous forests that enhances the fire events to 75% of the total (Negi and Dhyani 2012). Currently, more than 17.8% of Himachal Pradesh forests is covered with Chir pine trees, very vulnerable to forests fires. Forest fires in Uttarakhand are mostly associated with anthropogenic causes such as land use dynamics, particularly near roads and human settlements (Verma et al. 2019). The study conducted for part of Indian Western Himalayas comprising three states of Jammu and Kashmir, Himachal Pradesh and Uttarakhand (Pragya et al. 2023) suggests the incidence of forest fires is specific to the forest types, regulated majorly by elevation, temperature, moisture conditions, and by nearness to the settlements and the roads in region. The study further suggested that higher susceptibility of forest fires is associated with the subtropical coniferous forests and tropical moist deciduous forests owing to the dominance of the relatively more flammable Chir pine species.

Forest fires in the Hindu Kush Himalaya (HKH) are resulting into the degradation of forests in the region and becoming a serious ecological concern (Vachula et al. 2020; Zhang-Turpeinen et al. 2020). More than 600,000 active fires were recorded in the HKH region during the decade of January 2010 to May 2021, with highest number of 311,540 in March, the summer season, and 170,317 incidents in April due to increase in temperature by 0.2 °C (average of 2010–2020) and dry conditions (Bhattarai et al. 2022). The number sharply decreased to 11,000 fire incidents in December, the winter month. It is further noted that the number of forest fires in Nepal rose to around 250% in February, March and April in 2021 and to around 200% during the same period in Uttarakhand and Himachal Pradesh coinciding with deficiency in rainfall at 18% in 2019 and 20% in 2020 (Guhathakurta et al. 2020).

The forest fire events in India are largely (almost 95%) caused by humans, whether accidental or intentional, such as grazers, sparks from vehicles, hiking or picnicking activities, shifting cultivation and burning litter (pine needles) for a flush of fresh grasses (Mina et al. 2023). Forest fires in Western Himalayan region are persistent hazard that is further intensified by changing climate trends. The geographical area of Uttarakhand, Himachal Pradesh and Jammu and Kashmir is characterized by diverse forest types, steep terrain and a wide range of climatic conditions, making it particularly vulnerable to forest fires. Massive fires have encompassed the Nainital, Haldwani and Ramnagar forest divisions of Uttarakhand and forests of Himachal Pradesh during April–May 2024. The Forest Department of Himachal Pradesh reported a total of 1684 forest fires.

Himalayan region witnesses forest fires during the pre-monsoon summer period—a period of moisture stress due to resultant depletion of snowmelt water and high atmospheric temperature. The human activities, such as unattended campfire, discarded cigarettes, sparks from vehicle, shifting cultivation, etc., further compound the forest fire hazards.

The frequency and magnitude of mountain hazards is likely to increase, enhancing the risk of geo-hazards dramatically under the impact of climate change and in combination with increase in the population. The rising trend of extreme warm events, a falling trend of extreme cold events, a rising trend in extreme values and frequencies of both minimum and maximum temperature-based indices in the HKH for the past five to six decades will continue and pose even more acute challenges to adaptation (Krishnan et al. 2019). The lack of ubiquitous local weather stations coverage in the Indian Himalaya causes limitation in quantifying the extreme weather conditions to help understand the process that triggers the natural hazards. Nevertheless, the occurrences of cloudburst and associated mass wasting in a set-up of the closed valley with a narrow outlet in Himalayan terrain suggest that the extreme precipitation remains localized in the close catchments with high slopes and relief and often drained by streams with orthogonal aligned and high stream gradient (Dimri et al. 2017). The extreme/heavy precipitation events of flash floods, GLOFs, LLOFs, snow avalanches and landslides have rapidly increased in terms of frequency and magnitude due to the warming of climate in Himalayan region. The increase in precipitation extremes is well related to almost 77.52% per unit increase in temperature over India in last 113 years including the Himalaya with approximately 25% decrease in the snowfall influencing the snow depth amounting to 71 $\pm$ 24% decline in the region (Rafiq et al. 2022). Intensification of climate extremes is predicted from most of the climate models which may directly impact the sustainability of the region in future.

# Chapter 7
# Climate Change vis-á-vis Human Livelihood and Agriculture in NW Himalaya

**Abstract** The mountain agro-ecosystem is seriously affected by the population density in northwest Himalaya with 25.9 million people representing 48.5% of the total population of Himalaya. The people living in Himalaya are aware of the serious implications of depletion and threat to biological diversity considering the growing population and climate change. Changes in the availability of water due to altered winter and summer precipitation, decreasing winter snowfall and strengthening summer monsoon, show a pattern of reduced NDVI during the early growing season, and increased NDVI towards the end of the growing season. The increase in agriculture production in the state of Himachal Pradesh and the Union Territory of Jammu & Kashmir by 7% over the period from 1970 to 2004, is attributed to the improvement in the facilities of irrigation and introduction of cash crops in the cropping pattern. Crop diversification towards fruits and vegetables are becoming important components of the cropping pattern in the changing climate. The natural and/or anthropogenic degradation has implications on the food production, water resources and human health of the region, in general, and society, in particular. The projected changes in precipitation in Himalaya shall favour both, droughts and floods, in future with consequences on the food production.

**Keywords** Population density · Agro-ecosystem · Cropping intensity · Cropping pattern · NDVI (Normalized Difference Vegetation Index) · Land use land cover · Crop diversification · Socio-economic study

The water released from the glaciers, covering the Himalaya and Karakoram extensively, particularly during dry seasons, serves as lifeline to millions of people living in the mountain valley catchments of Indus, Ganges and Brahmaputra Rivers. The recent glacier cover estimates for the Himalaya and Karakoram are ~ 42,000 km$^2$, of which 48% is in the Karakoram, while 21, 23 and 8% is in the western, central and eastern Himalaya, respectively (Bolch et al. 2012; Nuimura et al. 2015). About 10% of total glacierized area is debris covered, mostly the Karakoram and southern slopes of the Himalaya (Kraaijenbrink et al. 2017). The glaciers in Himalaya-Karakoram are the primary local water resource for the sustenance of lowland rivers. The



R. K. Ganjoo and M. N. Koul, *The Himalayan Glaciers, Climate Change and Society*,
https://doi.org/10.1007/978-3-031-78283-1_7

highest contribution of ice/snow meltwater is during the spring, summer and early autumn. The meltwater is relatively more in the post-monsoon period (September–November) in eastern and central Himalaya. The need for glacier meltwater is more in Karakoram and northwest Himalaya, in the background of growing population, increased food production and rising affluence (Momblanch et al. 2019; Biemans et al. 2019) where summer precipitation is low, whereas it is less in the monsoon-dominated regions with abundant summer precipitation. Further, abundance of melt-water sustains the hydropower generation, agriculture and ecosystems in Hindu Kush Himalayan region.

Northwest Himalaya, a large mountainous region, encompasses the Indus River, along with watersheds of major rivers of Chenab, Ravi, Beas and Satluj, and extend from the Union Territory of Ladakh to the state of Uttarakhand. Based on altitudinal variation, lithology and tectonic evolution, the northwest Himalaya runs vertically from the mountainous terrain of Lesser, High and trans Himalaya witnessing different climate zones of tropical, temperate and cold arctic (cryosphere) as the region is characterized by altitudinal zones varying from 2000 m amsl to more than 6000 m amsl. The northwest Himalaya is populated with 25.59 million people representing 48.5% of the total population of Himalaya (Apollo 2017) having agriculture as primary livelihood option. The significant increase in the size of population has made mark on the population density causing serious effect on the mountain agro-ecosystem. In the present scenario of changing climate, the dense population of Himalaya are experiencing demographic, environmental and social stresses.

## 7.1 Culture, Lifestyle and Household—A Case Study of Ladakh Region

The cold desert region of Himalaya has a unique culture of its own, which is an amalgam of Indian and Tibetan. It is reflected in the Buddhism practised in Himalaya and manifested in the art, architecture, lifestyle, food habits, attire, etc.

Earliest inhabitants of Ladakh consisted of a mixed Indo-Aryan population of *Mons* and *Dards*. The physical features and dresses of the people of Ladakh are more akin to those of Tibet and Central Asia than to the mainland India. Buddhism came to western Ladakh from Kashmir in the second century when much of the eastern Ladakh and western Tibet was still practising *Bon* religion. Ladakh is also called 'Little Tibet' as it has been influenced by Tibetan culture. In the past, Ladakh gained importance from its strategic location at the crossroads of important trade routes, but after the borders with Tibet and Central Asia were closed by the Chinese authorities in 1960, international trade dwindled. Since 1974, the Indian government has encouraged tourism in Ladakh, as a major Himalayan tourist destination, which is a fast-emerging industry.

The Union Territory of Ladakh comprises the districts of Leh and Kargil and nearly 60% of the total population follow Buddhism. Buddhists are predominantly in the Leh district and Zanskar tehsil, whereas Shiite Muslims predominate in the Kargil district. Padum, the tehsil headquarter of Zanskar, is inhabited by Buddhists and Muslims. The composite population is descent of Tibeto-Dards. It is the only settlement of Sunni population believed to be descendent of people who came along with General Zorawar Singh from Kishtwar on an expedition to conquer Ladakh and Tibet. The victorious General Zorawar Singh granted land at Padum in Zanskar to the followers. The detailed socio-economic survey of the village households of Zanskar and Nubra, conducted during 2011–2013, reveal that there is a complete emancipation of women in the Buddhist society, giving them a special status in the society. Nearly 52% of the population is male, 42% is female and more than 46% of the population falls in the age group of 15–45 years. The population, in the age group of 15–45 years, is biologically reproductive, economically vibrant and demographically mobile. Society is vibrant and progressive in caring for livelihood and meeting the demands of entire society. The total literacy rate is nearly 20%, literacy rate among male is 75% and among female is 25%. Further, Zanskar and Nubra have a high sex ratio of 893 and 1147 females, respectively, per 1000 males in Buddhist predominant villages comprising 50% of the total villages, compared to 788 females per 1000 males at Padum (tehsil headquarter of Zanskar) and Diskit (Nubra valley) villages with composite population.

The age-old pastoralist, the nomads of Ladakh, sustained the harsh climate conditions to grant a peculiar political, social and cultural identity to the region. The influx of modernization in the region, in recent times, has improved the quality of living of the native population with the introduction of various technologies for the sustainable development. The Buddhists of Ladakh have preserved the culture trait in their home and lifestyle. The people of Ladakh build cozy houses to meet the demand and comfort during harsh chilly winters. The functional feature of the house is the lay and sheaves of dry lucerne used to border roofs up to half a metre. People of Ladakh consider the room which has a family hearth, called as *Thab-La-Gyamo* or Hearth Goddess, as main living room. The family members sit in this room around the central fireplace where food is cooked. The fireplace is made of special fireclay and the smoke is drawn out by pipes. Cow dung cakes and firewood are used for fuel. Thick mattress covered with carpets are used on the floor as well as walls for sitting. The basement room of the house is meant to preserve and store the raw vegetables with roots and raw dry pieces of meat which is consumed during the winter season. The *momos* (minced meat mixed with finely chopped onion, ginger and spices) is the delicacy of Ladakh cuisine. The tea lace with butter and served in a special container with *satoo* (powdered cereal), is considered favourable beverage in the households of Ladakh. The traditional attire of the men of Ladakh is warm woollen in white called *La-pa-sha*, coarse woollen legging around the legs, long woollen (thick *Patoo*) shoes and locally prepared skull cap. The women in Ladakh wear black-coloured woollen jacket, elegant *chuba* and broad-striped woollen lowers called *petticoat*. *Goncha*, the ethnic dress of Ladakh, is worn by all. Age-old ethnic dances, heritage from Tibet, is preserved in Buddhist culture of Ladakh. The principal dance is *kreteek-chenmo-shondeol*. The fascinating devil dance has spiritual significance and is performed by *Lamas* of *gompas*. People in Ladakh and Zanskar display the culture in gaiety and colour and celebrate many festivals in different parts of the region, such as *Losar Mela* (New Year festival), etc. The festivals of Ladakh region are the attraction for the tourists.

Almost all the Buddhist localities have monasteries known as *Gompa*, built on plain surface or atop the hillock neighbouring settlement, and decorated with paintings of Lord Buddha. *Lamas* (Buddhist priests) practise dancing, singing and playing on pipes and horns. The *Gompas*, centre of the cultural life of people, have influenced the religious beliefs for centuries. The *Gompas* play a dominant role in binding the society and help farmers in managing the farming. Besides, *Chorten* and *Mani* walls are also religious and architectural manifestations, unique to the region.

## 7.2 Mountain Agriculture, Climate and Diversification

The resources that are essential to agricultural economic growth in the hilly areas of Kashmir (Jammu and Kashmir), Himachal Pradesh and Uttarakhand, particularly at the watershed level, are threatened by rapid population growth, extreme poverty, loss of biodiversity, pollution of air and water, soil toxicity and erosion and implementation of short-sighted economic policies. As a consequence, sustainable development of hilly region has emerged as a key issue in the agricultural development planning. The mountain and hill states/union territories of northwest Himalaya comprising Jammu and Kashmir, Ladakh, Himachal Pradesh and parts of Uttarakhand and the Hindu Kush Himalayan (HKH) region presents a unique story of development, particularly with reference to Himachal Pradesh (India) that witnessed a revolution in the 1990s in social infrastructure and emerged as one of the most progressing states in north India compared to other developed mountain states (Negi et al. 2018).

Agricultural production systems can be sustainable if they are maintained without damage to regenerative capacity and diversity of natural resources. There is increased awareness among the mountain people on the serious implications of depletion and threat to biological diversity in the light of the growing population and climate change. In view of the factors of inaccessibility to remote areas of the mountain regions and fragile ecosystem, there is quest for ecologically sound agricultural systems and alternate development approaches to cater the demands of preservation to biodiversity and poverty alleviation.

Climate plays a pivotal role in the economic growth of hill states especially in the field of agriculture and horticulture. The habited mountainous regions at higher altitudes experience long, cold and chilly winter from the month of October to May, on the one hand, and short summers, on the other hand. People of the region have accordingly adapted themselves to the harsh cold arid climate and rugged terrain through ages by developing indigenous institutional set-ups and customs. Cultivated land cannot be extended easily to meet the need of households, as both, the cultivable land and water, are limited in mountainous regions. The land, in high-altitude mountainous (Ladakh, Zanskar, Lahaul-Spiti, Kinnaur, Uttarkashi Gangotri watershed) areas, has been made suitable for farming by levelling and terracing. Landholdings, in general, are small in size and large continuous stretches of cultivable land on mountain slopes are lacking. The past mountain development perspectives in the HKH region are found to be unable to capture existing 'socio-economic' and 'biophysical' changes, calling for the modernization of the current practices (Bhatta et al. 2019). In mountainous regions, the development programmes must be formulated in sync with the environment to benefit the mountain people.

Cropping intensity is highest (196–176%) in the Jammu division because of the suitable topography and tropical climate (Hussain 2020). In Uttarakhand, about 22.4% of the area is under cultivation and nearly 80% of people are dependent on it. A very limited agriculture is practised, primarily due to the nature of terrain in the state, and the net-sown area, therefore, is a little over 11% of the total land area (Khalid 2022). The net-sown area has decreased from 737,000 hectares to 680,000 hectares from 2010 to 2020, whereas the gross-irrigated area has decreased from 567,000 hectares in 2010 to 561,000 hectares in 2020 and the net-irrigated area has also decreased from 338,000 hectares to 326,000 hectares from 2010 to 2020 (Verma et al. 2022). The cultivation of crops in Leh, Kargil, Zanskar (Figs. 7.1 and 7.2), adjoining areas of Himachal Pradesh and Uttarakhand is supported wherever the soil moisture retention capacity of land is high enough whereas other parts of land are left uncultivated. About 75–85% of the population in Suru-Zanskar valley is engaged in age-old agro-pastoralist form of land use and agriculture is the mainstay of the people (Table 7.1). Changes in the availability of water due to altered winter and summer precipitation, decreasing winter snowfall and strengthening summer monsoon, show a pattern of reduced NDVI (Normalized Difference Vegetation Index) during the early growing season, and increased NDVI towards the end of the growing season. The low population density ($< 2$ person/km$^2$) of Ladakh region served an ideal factor, in as early as 1981, to conserve the Himalayan biodiversity with a result of setting up of first high-altitude Hemis National Park.

Agricultural implements, used by farmers of the region, are crude and primitive in form. Subsistent farming is still in practice because of the challenges of climate, steep foothills, fragmented small landholdings and terrain inaccessibility. The manure, until recent times, was added to the agriculture fields through a pool of cooperative farming. Most (85%) of the population in landlocked Zanskar and Nubra valley is engaged in sedentary cultivation and animal rearing. Agricultural implements, such as *Zho* (deployed in drawing the small plough) and wooden drum (used in churning milk to butter), though primitive are still in use. The land utilization pattern in Ladakh region, such as arable fields, pastures, grasslands, has increased considerably since 2001 under different natural habitats due to climate change (Table 7.2) in terms of marginal increase in mean monthly maximum temperature and rainfall during the summer season. The change has led to the growth of *yarkandi* grass, unique and representative of the Tibetan Plateau, in high alpine pasture zone (4000–4200 m amsl) that is harvested as fodder for livestock.

Families essentially rely on subsistence agriculture based on principal crops like wheat, barley and potato grown on the stone-built terraces. Potato is the main vegetable in the region and grown as the cash crop between the months of May and September. In last two decades, agriculture practices have changed in the villages of

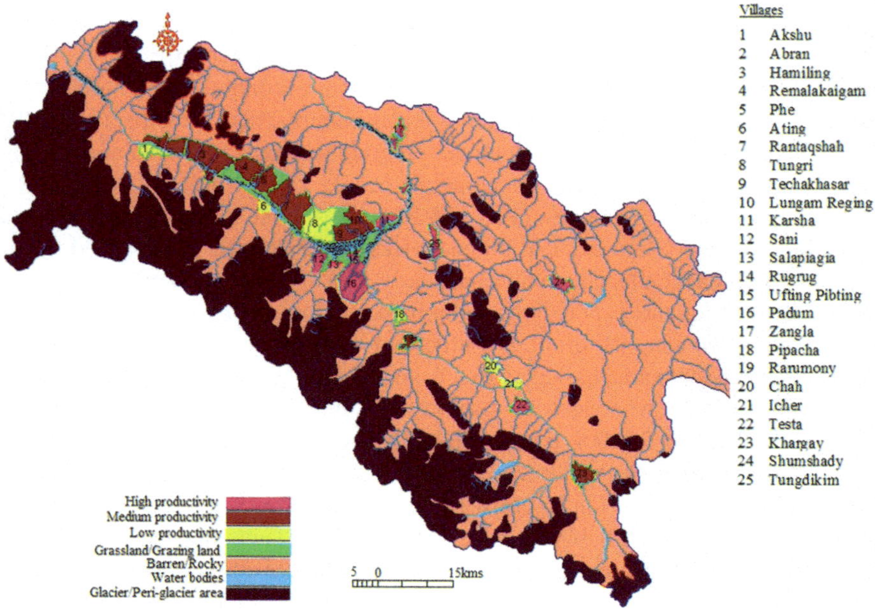


**Fig. 7.1** Agriculture productivity of villages in the cold desert of Suru-Zanskar valley, Ladakh

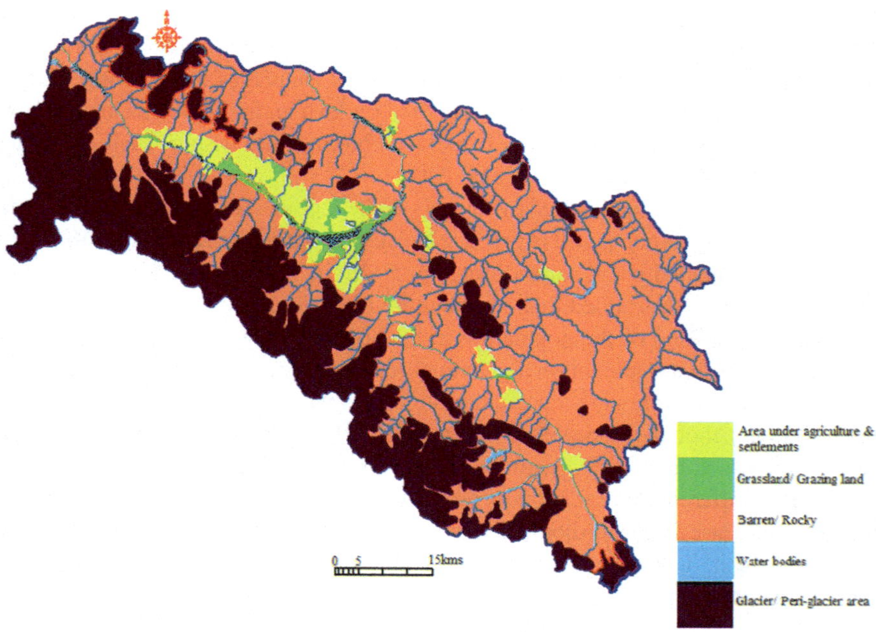


**Fig. 7.2** Land cover/use map of Suru-Zanskar valley, Ladakh

**Table 7.1** Agricultural productivity of villages located in Suru- Zanskar valley, Ladakh

| Villages | Yield wheat | A.W(C) | Index | Y.G (Y) | A.G (C) | Index | Y.P(Y) | A.P (C) | Index | Y.F(Y) | A.F (C) | Index | Y.V(Y) | A.V(C) | Index | E.I (agriculture) |
|---|---|---|---|---|---|---|---|---|---|---|---|---|---|---|---|---|
| Rugrug | 10 | 7.02 | 105.3 | 11 | 8 | 91.0 | 9.5 | 19.09 | 105.56 | 16.5 | 106.36 | 103 | 9.5 | 3.93 | 105.5 | 102.98 |
| Padum | 10 | 43 | 105.3 | 12 | 77 | 100 | 9.5 | 31.04 | 105.56 | 16.5 | 64.11 | 103 | 9.5 | 4 | 105.5 | 102.84 |
| Uftipibting | 10 | 29.93 | 105.3 | 12 | 65.31 | 100 | 9.5 | 21 | 105.56 | 16.5 | 84.26 | 103 | 9.5 | 3.8 | 105.5 | 102.73 |
| Shumshadi | 10 | 20 | 105.3 | 10 | 3.93 | 83.3 | 9.5 | 13.02 | 105.56 | 15 | 3.56 | 93.8 | 9 | 1.14 | 100 | 102.16 |
| Sani | 10 | 28.09 | 105.3 | 12 | 29 | 100 | 9 | 4.04 | 100.00 | 16 | 13.52 | 100 | 9.5 | 3 | 105.5 | 102.12 |
| Zangla | 10 | 43 | 105.3 | 12 | 40 | 100 | 9.5 | 35 | 105.56 | 15 | 22 | 93.8 | 9.5 | 2 | 105.5 | 102.07 |
| Tungdikim | 10 | 37 | 105.3 | 12 | 56.02 | 100 | 9.5 | 21.03 | 105.56 | 16 | 45.6 | 100 | 9 | 1.5 | 100 | 101.93 |
| Salapi | 10 | 14.5 | 105.3 | 12 | 31.9 | 100 | 9.5 | 32 | 105.56 | 16 | 61.42 | 100 | 9.5 | 1.98 | 105.5 | 101.87 |
| Karsha | 10 | 28.9 | 105.3 | 12 | 31.4 | 100 | 9 | 15.02 | 100 | 16 | 50.01 | 100 | 9.5 | 2.27 | 105.5 | 101.29 |
| Testa | 9.5 | 4 | 100 | 12 | 39.93 | 100 | 9.5 | 18.5 | 105.6 | 16 | 34.97 | 100 | 9 | 2.5 | 100 | 101.04 |
| Remalakia | 9.5 | 18 | 100 | 12 | 40 | 100 | 9 | 10 | 100 | 16.5 | 21.68 | 103 | 9 | 1.02 | 100 | 100.75 |
| Rarumoon | 9.5 | 4.1 | 100 | 11 | 22.5 | 91.7 | 9 | 9 | 100 | 17 | 23.58 | 106 | 9 | 1.17 | 100 | 99.335 |
| Kargay | 8 | 1.07 | 0 | 12 | 39 | 100 | 8 | 0.04 | 0 | 16 | 41.64 | 100 | 9 | 1.53 | 100 | 98.667 |
| Techa | 10 | 20 | 105.3 | 11 | 27.6 | 91.7 | 9.5 | 15.7 | 105.56 | 15 | 14.59 | 93.8 | 9 | 1.56 | 100 | 98.38 |
| Hamiling | 9 | 9 | 94.74 | 10 | 10.5 | 83.3 | 9.5 | 9.5 | 105.56 | 16.5 | 25.27 | 103 | 8 | 0.98 | 88.88 | 98.163 |
| Lungmireg | 10 | 9.5 | 105.03 | 11 | 28.2 | 91.7 | 9 | 17 | 100 | 16 | 27.47 | 100 | 9 | 1.53 | 100 | 97.79 |
| Abran | 9.5 | 12.04 | 100 | 10 | 17.09 | 83.3 | 9.5 | 19 | 105.56 | 16 | 17.88 | 100 | 9 | 1.09 | 100 | 97.328 |
| Phe | 9 | 8 | 94.74 | 11 | 30 | 91.7 | 9 | 14.5 | 100 | 16 | 51.1 | 100 | 9 | 2.5 | 100 | 97.247 |
| Rattaqshah | 9.9 | 14 | 104.02 | 10 | 14.17 | 83.3 | 9 | 9.5 | 100 | 16 | 23.3 | 100 | 9 | 1.15 | 100 | 97.147 |
| Ating | 9.5 | 12.97 | 100 | 11 | 28.4 | 91.7 | 9 | 16.3 | 100 | 16 | 14.78 | 100 | 8 | 0.5 | 88.88 | 96.68 |
| Pipcha | 9.5 | 2.14 | 100 | 10 | 8.12 | 83.3 | 9 | 2.9 | 100 | 16 | 17.86 | 100 | 9.5 | 1.93 | 105.5 | 96.218 |

(continued)

**Table 7.1** (continued)

| Villages | Yield wheat | A.W(C) | Index | Y.G (Y) | A.G (C) | Index | Y.P(Y) | A.P (C) | Index | Y.F(Y) | A.F (C) | Index | Y.V(Y) | A.V(C) | Index | E.I (agriculture) |
|---|---|---|---|---|---|---|---|---|---|---|---|---|---|---|---|---|
| Icher | 9.5 | 4.02 | 100 | 10 | 12 | 83.3 | 9 | 3 | 100 | 15.5 | 13.6 | 96.9 | 9 | 1.03 | 100 | 92.793 |
| Akshu | 9.5 | 3 | 100 | 10 | 23 | 83.3 | 8 | 9.5 | 88.889 | 16 | 14.5 | 100 | 9 | 1.25 | 100 | 90.461 |
| Chea | 9.5 | 5.15 | 100 | 10 | 16.12 | 83.3 | 8.5 | 2.07 | 0 | 16 | 19.31 | 100 | 8 | 0.7 | 88.88 | 88.848 |
| Tungri | 9.5 | 12.02 | 100 | 11 | 24 | 91.7 | 9 | 14 | 100 | 16.5 | 43.18 | 0 | 9 | 2 | 0 | 50.441 |

*Y* Yield *C* Crop *YW* Yield Wheat *YG* Yield Grim *YP* Yield Peas *YF* Yield Fodder *YV* Yield Vegetable *AW* Area Wheat *AG* Area Grim *AP* Area Peas *AF* Area Fodder *AV* Area Vegetables *EI* Efficiency Index (after Bhatia 1967)

**Table 7.2** Average consumption of grasses/ bushes and excreta released by the livestock population of Suru-Zanskar valley, Ladakh

| S. No. | Type of animals | Total livestock population (2001) | Average excreta released kgs/day/ animal head in wet form | Average excreta released kgs/day/ animal head in dry form | Total excreta released by Zanskari cattle population in dry form | | Per day consumption in Kgs/animal head | Total consumption of livestock population | |
|---|---|---|---|---|---|---|---|---|---|
| | | | | | in tones/ day | in tones /year | | in tones/ day | in tones/year |
| 1 | Cow | 6594 | 5.5 | 1.76 | 11.60544 | 4235.986 | 16.5 | 108.80 | 39,449.2 |
| 2 | Yak | 1577 | 6.4 | 2.04 | 3.21708 | 1174.234 | 25.86 | 40.78 | 14,884.7 |
| 3 | Dzo | 1223 | 6.4 | 2.04 | 2.49492 | 910.6458 | 25.86 | 31.62 | 11,541.3 |
| 4 | Dzomo | 813 | 6.4 | 2..04 | 1.65852 | 605.35 | 25.86 | 21.02 | 7672.3 |
| 5 | Sheep | 95,000 | 1.8 | 0.57 | 54.15 | 19,764.75 | 2.87 | 272.65 | 99,517.25 |
| 6 | Goat | 24,000 | 1.8 | 0.57 | 13.68 | 4993.2 | 2.87 | 68.88 | 25,141.2 |
| 7 | Other Specific Equines | 1.509 | 5.4 | 1.72 | 2.59548 | 947.3502 | 12.36 | 18.65 | 6807.25 |
| Total | | 130,716 | | | 89.40144 | 32,631.515 | | 562.4 | 205,013.2 |

Zanskar and Nubra valleys, that are now connected with roads, where mechanization and recent scientific agriculture techniques have been introduced. The valleys support cultivation of selected crops, such as wheat, grim (barley), pea, fodder and vegetables, that grow in a short period and per hectare output of the crops which has recently increased due to the use of hybrid variety of seeds suitable to the harsh climate conditions but with assured water supply. Further, in adjoining mountainous regions where the crop has the least feasibility, livestock rearing is a natural alternate supported by a more conducive sub-arid cold climate during summer seasons. The shortage of labour and short growing season have been efficiently managed by *gompas* (Buddhist monasteries) through the collective farming system. Cattle population in Zanskar valley is primarily managed by women pastoralists having training in dairying. The women pastoralists, because of transhumance skills move to high alpine pastureland settlement (locally called *Doksa*) to graze and rear cattle, and mulch during summer months (Table 7.3). Zanskar valley is famous to produce local butter (Fig. 7.3a and b), dry cheese, rearing and fleecing the variety of goats and *Yak* for *Pashmina* wool. The butter, cheese and *Pashmina* wool has large market in Kashmir for weaving very expensive *shawls* with a great demand in global market. Farmers are gradually shifting the focus from traditional cereal crops (subsistence farming) to high-value cash crops (commercial farming) such as fruits, vegetables, floriculture and medicinal plants.

Agriculture production in the state of Himachal Pradesh and the Union Territory of Jammu & Kashmir has increased by 7% over the period from 1970 to 2004. The increase is attributed to the improvement in the facilities of irrigation and introduction of cash crops in the cropping pattern. Although there is an increase in the irrigated land to 80% yet a large portion (77%) of the net-sown area continues to be rainfed. It is observed that the area available for cultivation has increased in Jammu and Kashmir from 675,000 hectares in 1966–1967 to 752,000 hectares in 2004–2005 and the net-sown area has increased from about 28 to 31% of the total area during the same period. At the same time, per capita availability of net-sown area has declined significantly, indicating growing pressure of the increasing population on the land and other resources of the state. The net-irrigated area has increased from 280,000 hectares to 311,000 hectares, but the irrigated area as a per cent of net-sown area has remained, more or less, the same over the period. The land under cultivation has been found highest in the districts of Budgam (>66%), followed by Kupwara (65%), and Anantnag (64%) in Kashmir division. Almost cent per cent of the net-sown area in Kargil district has irrigation facilities, followed by the districts of Leh and Srinagar. The land in the plains and intermontane basin of Jammu and Kashmir, under cultivation for food grain crops, has increased from 534,000 hectares in 1960–61 to 634,000 hectares in 2001 and further to 682,000 hectares in 2015–2016, indicating intensive cultivation due to the introduction of upgraded farm technology since 2001.

The food grain production in Jammu division has increased by 36% during the period from 1960 to 2015, whereas in higher altitudes the diversification has been initiated lately. Crop diversification in twelve districts of Himachal Pradesh were studied using Herfindahl–Hirschman's Index (HHI). The results indicated moderate diversification in eight districts—Shimla, Kullu, Sirmour, Lahaul-Spiti,

**Table 7.3** List of female herders in the selected *Doksas*

| Group head of *Doksas* | *Doksa*s head | Owner of villages | Age (years) | Cattle heads | | |
|---|---|---|---|---|---|---|
| | | | | Own | Others | Total |
| Shamkashi Yokma | 1. Dawa Tolma | Hongchet, Nyorek, Karsha | 63 | 3 | 27 | 30 |
| | 2. Dorjay Tolma | Renam, Karsha, Rezing, Phey | 68 | 4 | 22 | 26 |
| | 3. Tsewang Tolma | Karsha, Yulsum, Rantaksha | 49 | 4 | 26 | 30 |
| | 4. Kunzes Dolma | Karsha, Phey, Abran | 30 | 10 | 10 | 20 |
| | 5. Lobzang Dolma | Ufti, Pipiting, Karsha | 25 | 4 | 15 | 19 |
| | 6. Padma Lanzes | Nyerok, Youlang, Rantaksha | 53 | 3 | 18 | 21 |
| Shamkashi Kongma | 1. Thukzay | Yulsum, Karsha, Hamiling | 35 | 7 | 18 | 25 |
| | 2. Lobzang Standon | Tahan, Youlang, Hamiling | 20 | 6 | 14 | 20 |
| | 3. Skalzang Tolma | Rezing, Yulsum, Hamiling | 45 | 6 | 22 | 28 |
| | 4. Yangchen Dolma | Karsha, Rantaksha, Skyagam | 46 | 6 | 29 | 35 |
| Lato Marpo | 1. Tsering Yangdol | Tungri, Rantaksha | 47 | 4 | 26 | 30 |
| | 2. Stanzin Putith | Phey, Tungri, Rantaksha, Langmi | 21 | 7 | 13 | 20 |
| | 3. Yangchan Dolma | Salapi, Tungri, Phey, Abran | 67 | 7 | 23 | 30 |
| | 4. Lobzang Kunzom | Langmi, Karsha, Kishrak | 22 | 7 | 23 | 30 |
| Bho Thang | 1. Tsering Yangchay | Youlang, Karsha, Phey | 48 | 8 | 17 | 25 |
| | 2. Stanzin Yetok | Tungri, Rantaksha, Phey, Manda | 20 | 3 | 19 | 22 |
| | 3. Tsering Chodol | Hongchet, Rantaksha, Phey, Tungri, Manda | 22 | 2 | 27 | 29 |
| | 4. Tsering Dolma | Youlang, Karsha, Phey, Manda | 20 | 10 | 20 | 30 |
| | 5. Tsering Tolma | Karsha, Phey, Manda, Tungri | 25 | 8 | 22 | 30 |
| Chakdo Karpo | 1. Tashi Lamo | Karsha, Abran | 33 | 8 | 13 | 21 |

(continued)

**Table 7.3** (continued)

| Group head of *Doksas* | *Doksas* head | Owner of villages | Age (years) | Cattle heads | | |
|---|---|---|---|---|---|---|
| | | | | Own | Others | Total |
| | 2. Spalzes Angmo | Karsha, Abran, Rantaksha, Youlang, Tungri | 21 | 6 | 14 | 20 |
| | 3. Chunit Tolma | Karsha, Langmi, Tahan, Abran | 40 | 4 | 25 | 29 |
| | 4. Stanzin Thangpo | Karsha, Youlang, Yulsum, Khishrak, Abran | 31 | 4 | 16 | 20 |
| Oma Tangtse | 1. Yangdol Dolkar | Stongde, Karsha | 20 | 9 | 20 | 29 |
| | 2. Dechen Dolkar | Rezing, Karsha | 48 | 0 | 26 | 26 |
| | 3. Tsering Angmo Rihan | Ufti, Pipiting, Karsha | 60 | 8 | 20 | 28 |
| | 4. Tsering Dolma | Karsha | 59 | 7 | 15 | 22 |
| | 5. Tsering Angmo Kongma | Karsha | 60 | 4 | 16 | 20 |
| | 6. Yangchen Dolma | Stongde, Tungri | 53 | 8 | 18 | 26 |
| | 7. Padma Tsomo | Ufti, Karsha | 53 | 7 | 22 | 29 |
| | 8. Lobzang Yeton | Ufti, Karsha, Phey | 26 | 4 | 20 | 24 |
| Range | | | 20–68 | 0–10 | 10–29 | 19–35 |
| Overall mean | | | 39.68 | 5.74 | 19.87 | 25.61 |
| Total number | | | 31 | 178 | 616 | 794 |

a) b)

**Fig. 7.3** **a** Females in pair churning the milk to **b** extract butter out of the churned milk

Mandi, Kangra, Solan and Chamba, and least diversification in rest four districts—Kinnaur, Bilaspur, Una and Hamirpur. None of the districts in Himachal Pradesh showed high diversification or complete specialization (Dhanta and Negi 2018). In Garhwal Himalaya, cultivation of cash crops including fruits, off-season vegetables and medicinal plants, has replaced traditionally grown cereal crops. Cultivation of food crops like cereal crops and oilseeds are not gaining momentum in the state, while fruits and vegetables are becoming important components of the cropping pattern. The responsible factor for the change includes better road connectivity that improved the means of transportation, availability of favourable microclimatic niches, availability of new crop inputs such as, hybrid seeds, chemicals and fertilizers, emerging new markets, etc. The niches like organic farming, natural farming, cultivation and utilization of local underutilized crops, local landraces with special characteristics, cold-water fish (trout), etc., are exploited to improve the economic status of the hill people.

Himachal Pradesh has taken lead in the development of non-farm crops because of its geographical location and better connectivity with major markets at Delhi, the capital of India. The flower cultivation has been introduced in the Chabri watershed in the vicinity of Shimla hills (district Solan, Himachal Pradesh). The non-farm crops such as, black gram (*mash*) and horse gram (*kulth*), introduced at the state level has met grand success (Modgil et al. 2019). Application of scientific and technological know-how has brought development in the field of agriculture from the age-old agro-pastoralist form of land use in the cold desert region of Ladakh, Himachal Pradesh and Uttarakhand. Widespread cultivation of apples, *Chilgoza* (Pine nuts), and peas in Kinnaur district, Himachal Pradesh, due to the marginal change in climate, is a noteworthy example. Farmers in Kullu valley of Himachal Pradesh have, partly to completely, shifted from food crop farming to vegetable cultivation, a good sign for the future farm economy. Recent study (Saurabh et al. 2020) in Lahaul and Spiti district of Himachal Pradesh reveal significant change in the cropping pattern with traditional crops like potato, barley and black pea being increasingly replaced by green peas. Apple, yet another cash crop, is gaining popularity in Lahaul and Spiti due to perceptible change in climate conditions (Singh et al. 2022b).

Ladakh has minimal presence of cash crop as age-old historic agro-pastoralist form of land use stays around. Linking Leh (Union Territory of Ladakh) with Kashmir (Union Territory of Jammu & Kashmir) and Manali (Himachal Pradesh) by national highways has integrated Ladakh region with the outside world, inviting transition and modernization of the economy to the region. The flow of developments in Ladakh infused progress in agriculture, pastoralist, dairy farming and cottage industry. Further, improvement in education enhanced the literacy rate that helped the people to engage in non-primary activities leading to an increase in off-farm employment opportunities to overcome the shortages of manpower. The improvement in road connectivity led to crop diversification and cropping patterns. In 1971, 84.69% of total workers were employed in primary (agriculture work) sector as against 3.55% and 11.76% in the secondary (construction work) and tertiary (other economic services) sectors, respectively. There has been a significant shift from the primary to the tertiary sector over a period of four decades from 1971 to 2011, with a marginal change in the secondary sector. The primary sector constituted 27.21% of the workers (Wani et al. 2018, Dolker 2018). Increase in the level of market integration, further led to the development activities. Vegetable gardening has become one of the economic opportunities to drive cash for the women farmers as they earn more through round the year vegetable cultivation during cold-arid winter in trenches.

## 7.3 Mountain Land Use Change and Climate

Most of the studies, on land use changes, have been conducted in Nepal and Central Himalaya (India), whereas west Himalaya has received less attention. The major conclusions of the studies reveal that the expansion of agriculture has reduced the forest cover in mountainous states occupying Lesser and High Himalaya and effected the community forests the most. Some studies have reported an increase in forest area, while others have observed no significant change in land use (vegetation type) categories (Negi et al. 2017). Multi-temporal satellite data analysis of 26 years (1991–2016) suggests changes in land use and land cover in middle Himalayas of Himachal Pradesh (Thakur et al. 2018) that include increase in the agricultural, horticultural and other anthropogenic activities at the cost of decline in forest cover and grassland.

In Uttarakhand 63.41% of the total geographical area is under forest cover, and it varies from 71.55% in Nainital to 16.99% in the Udham Singh Nagar. The total forest land is categorized as reserve forest (69.86%), protected forest (26.01%) and unclassed forest (4.13%). The share of dense and open forest cover is 35.6% and 9.2% (ISFR 2023), respectively. The forest covers an area of 10.63% of the total geographical area of the union territories (combined) of Jammu and Kashmir and Ladakh. According to ISFR (2023) the union territories have 4,280.48 $km^2$ under Very Dense Forest (VDF), 8,612.36 $km^2$ under Moderately Dense Forest (MDF) and 10,719.05 $km^2$ under Open Forest (OF). Forest Cover in the union territories has increased by 370.89 $km^2$ as compared to the previous assessment reported in ISFR 2017. The reserved, protected and unclassed forests are 87.21%, 12.61% and 0.18% respectively of the recorded forest area in the union territories. The forest cover varies from 61.63% in Doda to 21.82% in Kishtwar. Ladakh has a forest cover area of merely 0.060% of the geographical area. The net cultivated land area in Kashmir division is 60%, whereas in Jammu division it is 21.3%, an increase of 2.3% from 1980–81 to 2005–2006. Land use-land cover analyses of Kashmir valley reveal that ~44% of the area is naturally vegetated comprising forests, scrub land and pastures, whereas agriculture, orchards, settlements and plantation by human cover ~25% of the area (Rather et al. 2016). The analyses further suggest degradation of forest cover over the period from 1972 to 2010 (38 years) due to illegal logging, and 18.22% of forest area is converted to sparse forest or scrub land. During the period of 38 years, there is decrease in croplands and plantation by 25% and 13%, respectively.

Further studies carried out in Uttarakhand, particularly in the Garhwal and Rudraprayag regions using satellite images of 1976, 1998 and 2014, reveal decrease in the area of dense forests from 55.24% (1069.51 $km^2$) in 1976 to 44.18% (855.39

$km^2$) in 2014. On the other hand, open forests increased from 19.18% (371.3 $km^2$) in 1976 to 23.82% (461.08 $km^2$) in 2014. The reduction in forest cover is caused due to fragmentation as a result of conversion of forest land for the purposes of construction of hydropower projects (62.93 hectares), roads (187.52 hectares) and other activities (299.08 hectares) in two phases with an overall annual rate of change in the forest cover of 0.22% in the period from 1976 to 1998 and 0.27% in the period from 1998 to 2014 (Batar et al. 2017). Correspondingly, area of pastureland decreased from 11.78% (227.98 $km^2$) in 1976 to 9.91% (191.78 $km^2$) in 2014. Agriculture and built-up areas increased progressively from 3.78% and 0.14% in 1976 to 8.02% and 0.62% in 2014, respectively.

## 7.4 Climate Change and Mountain Socio-Economy

Population in Indus basin depend on natural resources and lead an agro-pastoral life with cottage industry, casual employment and tourism-related activities as other sources of livelihood (Abbasi et al. 2017). On the other hand, the overall socio-economic condition in the higher elevation in Upper Ganga basin (district of Rudraprayag) is low, with 37% of the population below the poverty line.

The change in climate refers to the change, either due to natural variability or human interference, over an extended period causing degradation and alteration in the eco-system of the planet earth (Forster et al. 2020, Chiang et al. 2021). Decline in the population of several migratory and resident birds, including the globally threatened black-necked crane, *Grus nigricollis* in the wetlands of Ladakh, in past few decades, has invited attention to the main socio-ecological changes which are (Bagchi et al. 2017) (a) agricultural intensification, (b) draining of wetlands for irrigation, (c) intensive use of wetlands for livestock production and (d) proliferation of recreational tourism, coupled with altered precipitation and temperature regimes. Although the climate has changed in Ladakh in contemporary timescales yet no remarkable shift in the phenology in Himalayan wetlands is reported. The wetlands are not severely water-limited eco-regions compared to the wider landscape and may, therefore, show weaker trends and be more resilient to phenological shifts. Nevertheless, day length is considered important than temperature in causing the onset of budding and sprouting of plants (Raza et al. 2019) which becomes a significant reason for showing no phenological shift. Between 2002 and 2015 (Bagchi et al. 2017), no change in phenology is reported from Ladakh and similar trend is observed from the adjacent regions of Tibetan highlands. Although, the phenological patterns in the Himalaya are not understood well because of the abundant richness of the species in the region and lack of historical data yet decrease in the length of dry season in tropical Himalaya is likely to disrupt the herbivore populations during the main period of leaf expansion causing direct impact on the plant phenology (Chmura et al. 2019). The aerological system of river valleys supported by the flight ability helps desert locust to penetrate in Zhangmu district (Nyalam County) and in Jilong county (both in Tibet) up to an altitude of about 1700 m amsl with a small group of locusts above 4000 m amsl in July 2020 (Liu et al. 2021b). The distribution of locusts to higher altitudes is in response to warming climates particularly the requirement of specified number of degree-days (DD) above the threshold temperature of 14.2$^0$C (Tu et al. 2020 and references herein). The climatic, orographic and geological barriers of Himalaya acts handicap to colonize new territory or adapt the physiology to changing conditions of climate.

The natural and/or anthropogenic degradation has serious implications on the food production, water resources and human health of the region, in general, and society, in particular (Weiskopf et al. 2020). Besides, the climate change affects the socio-economy (Bhardwaj et al. 2018) and energy use (Forster et al. 2020) of the region. The impact of anthropogenic pressure on the forest wealth of the Himalayan region has serious implications on the people of the region who depend, directly or indirectly, on forest resources for daily energy requirements. The rain-fed dependent agriculture in Himalayan region is vulnerable to the climate and related changes in precipitation patterns (Pandey et al. 2018, Sabeerali and Ajayamohan 2018, Krishnan et al. 2019b) and the frequency of extreme events (Karmakar et al. 2017, Sharma et al. 2018c, Singh et al. 2019), affecting the crop yield and livestock with serious consequences on the mountain people of the region. Further it has been observed that climate change has created new opportunities for diversification of agriculture as individual crop has gained frontier area in Hindu Kush Himalaya at the cost of ecosystem services degradation due to increase in temperature and summer rainfall (Krishna Bahadur et al. 2022), elaborated in earlier chapters. Studies indicate that Himalaya is warming more than global average rate with increase in temperature during the winter and autumn than in summer, and increases are larger at higher elevations.

The people in Himalayan region use and consume fuel wood, because of the limited availability of alternate energy source, as the major biomass energy source (Singh et al. 2021c) causing a negative impact on the water balance of the Himalayan region (Dhakal et al. 2019). Besides, the impact includes deforestation and land degradation which, in turn, accelerates the change in climate of the region. The socio-economic study at Drass (Kargil, Ladakh) (Sudan et al. 2022) confirms awareness among the people on climate change (Table 7.4) and its impact on fuelwood energy. The people at Drass experienced increase in temperature, decrease in snowfall and uncertain precipitation (Table 7.5), indicators to climate change, over the years. The changes, according to the people of Drass, may affect the livelihood and energy resources (Table 7.6). The analyses of different climate datasets of Leh (Ladakh) region indicates a rapid increase in temperature and varied precipitation pattern in recent decades (Chevuturi et al. 2018) which is a forewarn to irreversible impacts over the land and people of the region leading to devastating consequences. The irreversible impacts include effect on the river flow particularly in lowland areas, recharge of the groundwater, composition of ecosystem, biodiversity, intensity and/ or frequency of natural hazards and the human livelihoods.

Panamik showed higher vulnerability (0.65) on the livelihood strategies as compared to Turtuk (0.50) and Diskit (0.43), in Nubra-Shyok valley (Ladakh). The fertile land at Turtuk and Diskit supporting the cultivation of different and

**Table 7.4** Awareness of climate change among respondents from Drass sub-basin, Kargil, Ladakh

| Land holdings | Yes (%) | No (%) | May be (%) |
|---|---|---|---|
| < 1 ha N = 323 | 253 (78.32) | 1 (0.309) | 69 (21.36) |
| > 1 ha N = 117 | 99 (84.61) | .... | 18 (15.38) |
| Total N = 440 | 352 (80.0) | 1 (0.22) | 87 (19.8) |

**Table 7.5** Change in climate experienced by respondents from Drass sub-basin, Kargil, Ladakh

| Land holdings | Changing rainfall pattern (%) | Rise in temperature (%) | Reduction in snowfall | |
|---|---|---|---|---|
| | | | Yes (%) | No (%) |
| < 1 ha N = 318 | 155 (48.74) | 163 (51.26) | 194 (61.01) | 124 (38.99) |
| > 1 ha N = 116 | 62 (53.45) | 54 (46.55) | 93 (80.17) | 23 (19.83) |
| Total N = 434 | 217 (50) | 217 (50.0) | 287 (66.12) | 147 (33.88) |

**Table 7.6** Usage of energy resources at Drass sub-basin, Kargil, Ladakh

| Land holdings | Alternate fuel other than fuelwood | | Collection of fuelwood | |
|---|---|---|---|---|
| | Animal dung (%) | LPG (%) | Yes (%) | No (%) |
| < 1 ha N = 323 | 323 (100) | 270 (83.6) | 80 (24.77) | 243 (75.23) |
| > 1 ha N = 117 | 117 (100) | 98 (83.76) | 44 (37.61) | 73 (62.39) |
| Total N = 440 | 440 (100) | 368 (83.63) | 124 (28.18) | 316 (71.82) |

introduction of new crops in comparison to Panamik, could be the cause for it (Tashi and Sudan 2022). The Livelihood Vulnerability Index (LVI) study (Tashi and Sudan 2022) further indicate that agriculture is not a primary source of livelihood in Panamik. Study also shows that Diskit and Turtuk have very low average crop diversity index of 0.19 and 0.13, respectively indicating cultivation of more variety of crops compared to Panamik with average crop diversity index of 0.27. The result of both indices, Livelihood Vulnerability Index (LVI) and Livelihood Vulnerability Index-Intergovernmental Panel on Climate Change (LVI- IPCC), point at Panamik as most vulnerable to climate change impacts in Nubra valley, while Diskit is least vulnerable because of the availability of better livelihood options. The rapid urbanization and boom in the tourism sector have impacted the livelihood of the people of Leh (Ladakh). The percentage of cultivators has gradually declined from 65.48% in 1971 to 28.02% in 2011 along with decrease in agricultural labourers from 7.36% to 1.41% during the same period (Bhattacharjee and Dutta 2023). The local people have diversified to other economic activities, such as tourism economy, brick making activities, jobs outside the Ladakh, etc. effecting the traditional agriculture.

## 7.5 Mountain Ecosystem and Future

Increase in the future precipitation is projected in the upstream basins of the Indus-Ganga–Brahmaputra (IGB) (Palazzi et al. 2015; Lutz et al. 2014) whereas precipitation is projected to decrease (Lutz et al. 2016) in parts of the HKH region (north-western part of the upstream Indus basin). The projected precipitation changes are in favour of both, droughts and floods, in future thereby having consequences for the food production. It is projected that precipitation extremes will increase mostly in the Upper Indus Basin with relative increases up to 130%, while temperature extremes are expected to be more frequent in Upper Brahmaputra Basin in the future (Wijngaard et al. 2017). Decreasing trend in monsoonal and annual rainfall is noticed in eastern part of IGB while increasing in the western part (Patel et al. 2021b). The streamflow in the upper Indus basin is dominated by the glacier meltwater (~41%), snowmelt (~22%) and rainfall-runoff (~27%) of the total runoff, whereas the streamflow in the upper Ganga and Brahmaputra River basins is mostly by the rainfall-runoff (66% and 59%) and the meltwater contributes (20% and 25%) to the total runoff respectively (Shrestha et al. 2015). The warming of region is likely to change the total runoff by − 5% to + 12% for Upper Indus, with increase by 1–27% for Upper Ganga and 0–13% for the Upper Brahmaputra basins by 2050 (Lutz et al. 2014). The hydrological regime of the region shall witness change in terms of changes in rainfall amount, an extreme event caused by rainfall, seasonal rainfall onset changes, rainfall intensity and duration due to significant change in precipitation pattern and rising temperature (Berger et al. 2019).

The traditional nutritious food crops, e.g. buckwheat, barley, millets, sorghum, oat and beans are disappearing from food system in HKH indicating decline in agro-biodiversity (Adhikari et al. 2019). The change in priority for productivity and degradation of grasslands due to climate has taken the toll of local breeds of livestock and poultry with decline in the population of yak in India, Nepal and Bhutan (Wu et al. 2016). Increased cultivation of popular cereals (e.g., wheat, maize and rice) and cash crops (Rasul et al. 2018) shows a shift from traditional crops to cash leading to an excessive use of inorganic inputs in mountain agriculture inviting serious challenges to the environment and sustainability (Hussain and Qamar 2020).

The human society is not linearly related to climate change. Instead, the change in the landscape and available resources of the region, due to climate change, prompt the humans to adopt to flexible and opportunistic strategies to cope with climate oscillations (Zerboni et al. 2016). The mid-Holocene transhumance occupational sites reported from Ladakh are indicators to the invasion of early man coinciding with climatic amelioration (Ganjoo 2015) and adaptation to the changing climate. Innumerable archaeological evidence point that human societies did not abandon the area affected by climate change but embraced new settlement strategies to meet emerging socio-economic challenges (Latorre et al. 2016 and references herein).

Livelihoods in the HKH region are mainly dependent on natural resources and subsistence agriculture, characterized by low yields due to limited access to productive resources and finance. Agriculture sector in Indus basin is vulnerable to floods and drought whereas horticultural sector in Upper Ganga is vulnerable to decrease in snowfall as required chilling is lacking for crops like peaches, apricots, plums and apples (Dilshad et al. 2019). More crop pest attacks on old varieties or poor yield due to an extreme maximum temperature or extreme events (drought, floods, etc.) which affected the growth negatively in the region made the farmers to employ change in the crop variety and diversify to cash crops. The growing recognition of the products, market access, remittance flow and expansion of cooperatives has encouraged the farmers of the mountain region of HKH to shift, over a period, towards subsistence of high value agriculture (Gioli et al. 2018). Based on high resolution climate model, the increase in global temperature at $2^0$C than at $1.5^0$C is going to enhance the vulnerability to food insecurity by increase in events of floods and droughts (Betts et al. 2018). The tropics may be an exception in the projected simulation with increase by 30% to 110% in the flows of Ganga while flows in Amazon could either increase by 3% or decrease by 25%.

The rise in temperature shall be the major driver in the retreat of glaciers leading to wide exposure of ice-free proglacial areas as a critical component of mountain and high-latitude landscapes (Rounce et al. 2023). The bedrocks or glacial sediments of proglacial landscape shall support a novel ecosystem of biotic colonization influencing the successive ecological communities (Pothula and Adams 2022) through the key processes of habitat filtering, biotic interactions and processes mediated by time (Hanusch et al. 2022). Both, direct and indirect, effects on the richness and composition of biotic community shall be realized over a period after the vacation of area by the glaciers (Ficetola et al. 2024).

# References

Abbasi, S.S., B. Ahmad, M. Ali, M.Z. Anwar, Z.H. Dahri, N. Habib, A. Hussain, B. Iqbal, S. Ishaq, N. Mustafa, R. Naz, Z.T. Virk, and P. Wester. 2017. The Indus basin: A glacier-fed lifeline for Pakistan. HI-AWARE Working Paper 11, 42. Himalayan Adaptation, Water and Resilience (HI-AWARE) Research Consortium. Kathmandu, Nepal.

Abdullah, T., S.A. Romshoo, and I. Irfan Rashid. 2020. The satellite observed glacier mass changes over the Upper Indus Basin during 2000–2012. *Scientific Reports* 10: 14285.

Adhikari, L., S. Tuladhar, A. Hussain, and K. Aryal. 2019. Are traditional food crops really 'Future Smart Foods?' A Sustainability Perspective. *Sustainability* 11 (19): 5236.

Adhikari, T.R., B. Baniya, Q. Tang, R. Talchabhadel, M.R. Gouli, B.R. Budhathoki, and R.P. Awasthi. 2023. Evaluation of post extreme floods in high mountain region: A case study of the Melamchi flood 2021 at the Koshi River Basin in Nepal. *Natural Hazards Research* 3: 437–446.

Agarwal, V., T. Bolch, T.H. Syed, T. Pieczonka, T. Strozzi, and R. Nagaich. 2017. Area and mass changes of Siachen glacier (East Karakoram). *Journal of Glaciology* 63 (237): 148–163.

Ahmad, I., and R. Chandra. 2018. Paleoenvironmental reconstructions of the late Quaternary loess-paleosol sediments of Kashmir valley. *Journal of Applied Geochemistry* 20: 59–90.

Ahmad, S.T., R. Ahmed, G.F. Wani, P. Sharma, and P. Ahmed. 2022. Glacier changes in Sindh basin (1990–2018) of north-western Himalayas using earth observational data. *Modeling Earth Systems and Environment* 8: 2567–2579.

Ahsan, S., M.S. Bhat, A. Alam, F. Hakim, and H.A. Sheikh. 2022. Evaluating the impact of climate change on extreme temperature and precipitation events over Kashmir Himalaya. *Climate Dynamics* 58: 1651–1669.

Ajai. 2018. Inventory and monitoring of snow and glaciers of the Himalaya using space data. In *Science and Geopolitics of the White World* eds. P. Goel, R., Ravindra, and S. Chattopadhyay, 5–113. Ahmedabad: Springer International Publishing AG.

Ali, I., A. Shukla, and S.A. Romshoo. 2017. Assessing linkages between spatial facies changes and dimensional variations of glaciers in the upper Indus Basin, western Himalaya. *Geomorphology* 284: 115–129.

Ali, S.N., J. Dubey, R. Ghosh, M.F. Quamar, A. Sharma, P. Morthekai, A.P. Dimri, M. Shekhar, M. Arif, and S. Agrawal. 2018. High frequency abrupt shifts in the Indian summer monsoon since Younger Dryas in the Himalaya. *Scientific Reports* 8: 9287.

Ali, S.N., S. Agrawal, A. Sharma, B. Phartiyal, P. Morthekai, P. Govil, R. Bhushan, S. Farooqui, P.S. Jena, and A. Shivam. 2020. Holocene hydroclimatic variability in the Zanskar valley, northwestern Himalaya, India. *Quaternary Research* 97: 140–156.





R. K. Ganjoo and M. N. Koul, *The Himalayan Glaciers, Climate Change and Society*, https://doi.org/10.1007/978-3-031-78283-1

Ali, S.N., P. Singh, P. Arora, P. Bisht, and P.P. Mortheka. 2022. Luminescence dating of late Pleistocene glacial and glacio-fluvial sediments in the Central Himalaya India. *Quaternary Science Reviews* 284: 107464.

Allen, S.K., P. Rastner, M. Arora, C. Huggel, and M. Stoffel. 2015. Lake outburst and debris flow disaster at Kedarnath, June 2013: Hydrometeorological triggering and topographic predisposition. *Landslides* 13: 1479–1491.

Alley, R.B., K.M. Cuffey, and L.K. Zoet. 2019. Glacial erosion: Status and outlook. *Annals of Glaciology* 60 (80): 1–13.

An, W., X. Hu, E. Garzanti, J.G Wang, Q. Liu. 2021. New precise dating of the India-Asia collision in the Tibetan Himalaya at 61 Ma. *Geophysical Research Letters* 48, e2020GL090641.

Anderson, L.S., W.H. Armstrong, R.S. Anderson, D. Scherler, and E. Petersen. 2021. The causes of debris-covered glacier thinning: Evidence for the importance of ice dynamics from Kennicott glacier, Alaska. *Frontiers of Earth Science* 723: 1–19.

Angchuk, T., A. Ramanathan, I.M. Bahuguna, A. Mandal, M. Soheb, V.B. Singh, S. Mishra, and S. Vatsal. 2021. Annual and seasonal glaciological mass balance of Patsio Glacier, western Himalaya (India) from 2010 to 2017. *Journal of Glaciology* 67 (266): 1137–1146.

Anoop, A., S. Prasad, B. Plessen, B. Basavaiah, G.R. Naumann, P. Menzel, S. Weise, and A. Brauer. 2013. Palaeoenvironmental implications of evaporative gaylussite crystals from Lonar lake, central India. *Journal of Quaternary Science* 28 (4): 349–359.

Antoniazza, G., and S.N. Lane. 2021. Sediment yield over glacial cycles: A conceptual model. *Progress in Physical Geography* 45 (6): 842–865.

Ao, H., E.J. Rohling, X. Li, Y. Song, A.P. Roberts, Y. Han, C.J. Poulsen, T.N. Jonell, D. Liebrand, Q. Sun, X. Li, X. Qiang, P. Zhang, and M.J. Dekkers. 2023. Northern hemisphere ice sheet expansion intensified Asian aridification and the winter monsoon across the mid-Pleistocene transition. *Nature Earth and Environment* 4: 36.

Apollo, M. 2017. The population of Himalayan regions by the numbers: Past, present and future. In *Contemporary Studies in Environment and Tourism,* eds. R. Efe, and M. Öztürk,, 145–160. Cambridge Scholars Publishing.

Ashraf, A., R. Naz, and M.B. Iqbal. 2017. Altitudinal dynamics of glacial lakes under changing climate in the Hindu Kush, Karakoram, and Himalaya ranges. *Geomorphology* 283: 72–79.

Ashutosh, S., S. Kumar, A. Choudhary, T. Biswas, and S. Ghosh. 2019. *Identification of fire prone forest areas based on GIS analysis of archived forest fire points detected in the last thirteen years.* Technical Information Series volume 1 number 1. Dehradun: Forest Survey of India, Ministry of Environment, Forest & Climate Change, Government of India.

Azam, M.F., P. Wagnon, E. Berthier, C. Vincent, K. Fujita, and J.S. Kargel. 2018. Review of the status and mass changes of Himalayan-Karakoram glaciers. *Journal of Glaciology* 64 (243): 61–74.

Azam, M.F., J.S. Kargel, J.M. Shea, S. Nepal, U.K. Haritashya, S. Srivastava, F. Maussion, N. Qazi, Chevallier, A.P. Dimri, A.V. Kulkarni, J.G. Cogley, and I.M. Bahuguna, 2021. Glaciohydrology of the Himalaya-Karakoram. *Science* 373 (6557), eabf3668.

Bae, S.W., K.E. Lee, T. Ko, R. Kim, and Y.-G. Park. 2022. Holocene centennial variability in sea surface temperature and linkage with solar irradiance. *Nature* 12: 15046.

Bagchi, S., E. Gupta, K. Murthy, and N.J. Singh, 2017. Assessing the evidence for changes in vegetation phenology in high-altitude wetlands of Ladakh (2002–2015). In *Bird Migration Across the Himalaya: Wetland Functioning Amidst Mountains and Glaciers, eds.* H. Herbert, T., Brins and T. Namgail, 205–216. Cambridge: Cambridge University Press.

Bahuguna, I.M., B.P. Rathore, R. Brahmbhatt, M. Sharma, S. Dhar, S.S. Randhawa, K. Kumar, S.A. Romshoo, R.D. Shah, R.K. Ganjoo, and Ajai 2014. Are the Himalayan glaciers retreating? *Current Science* 106 (7), 1008–1015.

Bahuguna, I.M., B.P. Rathore, A.S. Jasrotia, S.S. Randhawa, S.S.S. Yadav, S. Ali, N. Gautam, J. Poddar, M. Srigyan, A. Dhanade, P. Joshi, S.K. Singh, D.R. Rajak, and S. Sharma. 2021. Recent glacier area changes in Himalaya-Karakoram and the impact of latitudinal variation. *Current Science* 121 (7): 929–940.

Ballesteros-Cánovas, J.A., D. Trappmann, J. Madrigal-González, N. Eckert, and M. Stoffel. 2018. Climate warming enhances snow avalanche risk in the Western Himalayas. *Proceedings National Academy Science USA* 115: 3410–3415.

Ballesteros-Cánovas, J.A., T. Koul, A. Bashir, J.M.B. Del Pozo, S. Allen, S. Guillet, and M. Stoffel. 2020. Recent flood hazards in Kashmir put into context with millennium-long historical and tree-ring records. *Science Total Environment* 722: 137875.

Bandyopadhyay, D., G. Singh, and A.V. Kulkarni, 2019. Glacier elevation and mass changes in Himalayas during 2000–2014. *The Cryosphere* 1–16.

Banerjee, A., and A.P. Dimri. 2019. Comparative analysis of two rainfall retrieval algorithms during extreme rainfall event: A case study on cloudburst, 2010 over Ladakh (Leh), Jammu and Kashmir. *Natural Hazards* 97: 1357–1374.

Bar, S., B.R. Parida, G. Roberts, A.C. Pandey, P. Acharya, and J. Dash. 2021. Spatio-temporal characterization of landscape fire in relation to anthropogenic activity and climatic variability over the Western Himalaya India. *GIScience and Remote Sensing* 58 (2): 281–299.

Bar, S., B.R. Parida, A.C. Pandey, B.U. Shankar, P. Kumar, S.K. Panda, and M.D. Behera. 2023. Modeling and prediction of fire occurrences along an elevational gradient in Western Himalayas. *Applied Geography* 151: 102867.

Barnard, P.L., L.A. Owen, M.C. Sharma, and R.C. Finkel. 2004. Late Quaternary (Holocene) landscape evolution of a monsoon-influenced high Himalayan valley, Gori Ganga, Nanda Devi, NE Garhwal. *Geomorphology* 61: 91–110.

Basavaiah, N., E. Appel, B.V. Lakshmi, K. Deenadayalan, K.V.V. Satyanarayana, S. Misra, N. Juyal, and M.A. Malik. 2010. Revised magnetostratigraphy and characteristics of the fluviolacustrine sedimentation of the Kashmir basin, India, during Pliocene-Pleistocene. *Journal of Geophysical Research Solid Earth* 115: B08105.

Batar, A.K., T. Watanabe, and A. Kumar. 2017. Assessment of land-use/land-cover change and forest fragmentation in Garhwal Himalayan region of India. *Environments* 4 (2): 34.

Batbaatar, J., A.R. Gillespie, D. Fink, A. Matmon, and T. Fujioka. 2018. Asynchronous glaciations in arid continental climate. *Quaternary Science Reviews* 182: 1–19.

Beaud, F., J.G. Venditti, G.E. Flowers, and M. Koppes. 2018. Excavation of subglacial bedrock channels by seasonal meltwater flow. *Earth Surface Processes and Landforms* 43: 1960–1972.

Benn, D.I., and L.A. Owen. 1998. The role of the Indian summer monsoon and the midlatitude Westerlies in Himalayan glaciation: Review and speculative discussion. *Journal of the Geological Society of India* 155: 353–364.

Bera, A., B.P. Mukhopadhyay, and D. Das. 2019. Landslide hazard zonation mapping using multicriteria analysis with the help of GIS techniques: A case study from Eastern Himalayas, Namchi, South Sikkim. *Natural Hazards* 96: 935–959.

Berger, S., J. Bliefernicht, A. Linstädter, K. Canak, S. Guug, D. Heinzeller, L. Hingerl, M. Mauder, F. Neidl, E. Quansah, S. Salack, R. Steinbrecher, and H. Kunstmann. 2019. The impact of rain events on $CO_2$ emissions from contrasting land use systems in semi-arid West African savannas. *Science of The Total Environment* 647, 1478–1489.

Berkelhammer, M., A. Sinha, L. Stott, H. Cheng, F.S.R. Pausata, and K. Yoshimura. 2012. An abrupt shift in the Indian monsoon 4000years ago. In *Climates, Landscapes, and Civilizations,* eds. L. Giosan, D.Q. Fuller, K. Nicoll, R.K. Flad, and P.D. Clift, 75–87. Geophysical Monograph Series 198, American Geophysical Union.

Bernard, M., P. Steer, K. Gallagher, and D.L. Egholm. 2020. Modelling the effects of ice transport and sediment sources on the form of detrital thermochronological age probability distributions from glacial settings. *Earth Surface Dynamics* 8: 931–953.

Betts, R.A., L. Alfieri, C. Bradshaw, J. Caesar, L. Feyen, P. Friedlingstein, L. Gohar, A. Koutroulis, K. Lewis, C. Morfopoulos, L. Papadimitriou, K.J. Richardson, I. Tsanis, and K. Wyser. 2018. Changes in climate extremes, fresh water availability and vulnerability to food insecurity projected at 1.5°C and 2°C global warming with a higher-resolution global climate model. *Philosophical Transactions Royal Society A* 376: 20160452.

Bhambri, R., K. Hewitt, P. Kawishwar, and B. Pratap. 2017. Surge-type and surge-modified glaciers in the Karakoram. *Scientific Reports* 7 (1): 15391.

Bhambri, R., K. Hewitt, P. Kawishwar, A. Kumar, A. Verma, Snehmani, S. Tiwari and A. Misra. 2019. Ice-dams, outburst floods, and movement heterogeneity of glaciers, Karakoram. *Global and Planetary Change* 180: 100–116.

Bhambri, R., C.S. Watson, K. Hewitt, U.K. Haritashya, J.S. Kargel, A.P. Shahi, P. Chand, A. Kumar, A. Verma, and H. Govil. 2020. The hazardous 2017–2019 surge and river damming by Shispare glacier, Karakoram. *Scientific Reports* 10: 4685.

Bhardwaj, S.K., R.K. Aggarwal, and T. Kapoor. 2018. Energy consumption pattern in different agro-climatic zones in rural habitations of western Himalayan region, India. *Current World Environment* 13 (3): 380–389.

Bhargava, O.N. 2015. Evolution of the Tethyan and Karewa succession in Kashmir: A synthesis. *Journal of the Palaeontological Society of India* 60 (1): 51–72.

Bhargawa, A., and A.K. Singh. 2019. Solar irradiance, climatic indicators and climate change—An empirical analysis. *Advances in Space Research* 64 (1): 271–277.

Bhat, I.M., and T. Ahmad. 2021. Evolution of the Ladakh magmatic arc. *Journal of the Geological Society of India* 97: 980–984.

Bhat, I.M., T. Ahmad, and D.V. Subba Rao. 2019. Origin and evolution of Suru Valley ophiolite peridotite slice along Indus suture zone, Ladakh Himalaya, India: Implications on melt–rock interaction in a subduction zone environment. *Chemie der Erde Geochem.* 79: 78–93.

Bhat, I.M., H. Chauhan, R.A. Dar, and T. Ahmad. 2023. Ladakh Himalayan ophiolites (LHO): A geological heritage of northwestern India. *Geoheritage* 15: 2–17.

Bhat, I.M., T. Ahmad, D.V. Subba Rao, S. Balakrishnan, and N.V.C. Rao. 2021. PGE and isotopic characteristics of Shergol and Suru valley ophiolites, western Ladakh: Implications for supra-subduction tectonics along Indus Suture Zone. *Geoscience Frontiers* 12: 101118.

Bhat, S., and Q. Shah. 2017. Disaster profile of India: A theoretical understanding of regional and seasonal perspective. *IRA-International Journal of Management & Social Sciences* 6 (1): 52–59.

Bhatta, K.P., H.J.B. Birks, J.-A. Grytnes, and O.R. Vetaas. 2019. Weighted average regression and environmental calibration as a tool for quantifying climate-driven changes in vegetation. *Journal of Plant Ecology* 12 (3): 460–473.

Bhattacharya, G., D.M. Robinson, and M.M. Wielicki. 2020a. Detrital zircon provenance of the Indus Group, Ladakh, NW India: Implications for the timing of India-Asia collision and other syn-orogenic processes. *Geological Society of America Bulletin* 133 (5–6): 1007–1020.

Bhattacharya, G., D.M. Robinson, D.A. Orme, Y. Najman, and A. Carter, 2020b. Low-temperature thermochronology of the Indus Basin in central Ladakh, northwest India: Implications of Miocene-Pliocene cooling in the India-Asia collision zone. *Tectonics* 39: e2020TC006333.

Bhattacharya, A., K. Mukherjee, O. King, S. Karmakar, S.N. Remya, A.V. Kulkarni, J. Kropáček, and T. Bolch. 2023. Influence of climate and non-climatic attributes on declining glacier mass budget and surging in Alaknanda Basin and its surroundings. *Global and Planetary Changes* 230: 104260.

Bhattacharya, A., T. Bolch, K. Mukherjee, T. Pieczonka, J. Kropacek, and M.F. Buchroithner. 2016. Overall recession and mass budget of Gangotri glacier, Garhwal Himalayas, from 1965 to 2015 using remote sensing data. *Journal of Glaciology* 62 (236): 1115–1133.

Bhattacharjee, D. and R.K. Dutta. 2023. Mountainous hazards and sustainable livelihood in Leh valley. In *Indigenous Knowledge and Disaster Risk Reduction: Insight Towards Perception, Response, Adaptation and Sustainability* , eds. G.K. Panda, U. Chatterjee, N. Bandyopadhyay, M.D. Setiawati, and D. Banerjee, 83–105. Springer.

Bhattarai, N., S. Dahal, S. Thapa, S. Pradhananga, B.S. Karky, R.S. Rawat, K. Windhorst, T. Watanabe, R.B. Thapa, and R. Avatar. 2022. Forest fire in the Hindu Kush Himalayas: A major challenge for climate action. *Journal of Forest and Livelihood* 21(1): 14–31.

Bhushan, S., T.H. Syed, A.A. Arendt, A.V. Kulkarni, and D. Sinha. 2018. Assessing controls on mass budget and surface velocity variations of glaciers in Western Himalaya. *Scientific Reports* 8: 8885.

Bhutani, R., K. Pande, and T.R. Venkatesan. 2009. $^{40}Ar/^{39}Ar$ dating of volcanic rocks of the Shyok suture zone in north-west trans-Himalaya: Implications for the post-collision evolution of the Shyok suture zone. *Journal of Asian Earth Sciences* 34 (2): 168–177.

Bian, W., T. Yang, W. Peng, S. Wang, F. Gao, S. Zhang, H. Wu, H. Li, L. Cao, T. Jiang, and H. Wang. 2021. Paleomagnetic constraints on the India–Asia collision and the size of Greater India, *Journal of Geophysical Research: Solid Earth* 126 (6): e2021JB021965.

Biemans, H., C. Siderius, A.F. Lutz, S. Nepal, B. Ahmad, T. Hassan, W. von Bloh, R.R. Wijngaard, P. Wester, A.B. Shrestha, and W.W. Immerzeel. 2019. Importance of snow and glacier meltwater for agriculture on the Indo-Gangetic plain. *Nature Sustainability* 2: 594–601.

Bilham, R. 2019. Himalayan earthquakes: A review of historical seismicity and early 21st century slip potential. In *Himalayan Tectonics: A Modern Synthesis* eds. P.J. Treloar and M. P. Searle, 423–482. Geological Society, London: Special Publication 483.

Blöthe, J.H., H. Munack, O. Korup, A. Fülling, E. Garzanti, A. Resentini, and P.W. Kubik. 2014. Late Quaternary valley infill and dissection in the Indus River, western Tibetan Plateau margin. *Quaternary Science Reviews* 94: 102–119.

Bolch, T., M.F. Buchroithner, T. Pieczonka, and A. Kunert. 2008. Planimetric and volumetric glacier changes in the Khumbu Himal, Nepal, since 1962 using Corona, Landsat TM and ASTER data. *Journal of Glaciology* 54: 592–600.

Bolch, T., A. Kulkarni, A. Kääb, C. Huggel, F. Paul, J.G. Cogley, H. Frey, J.S. Kargel, K. Fujita, M. Scheel, S. Bajracharya, and M. Stoffel. 2012. The state and fate of Himalayan glaciers. *Science* 336: 310–314.

Bolch, T., T. Pieczonka, K. Mukherjee, and J. Shea. 2017. Glaciers in the Hunza catchment (Karakoram) have been nearly in balance since the 1970s. *The Cryosphere* 11: 531–539.

Bolch, T., J.M. Shea, and S. Liu. 2019 Status and change of the cryosphere in the extended Hindu Kush Himalaya Region. In *The Hindu Kush Himalaya Assessment: Mountains, Climate Change, Sustainability and People,* eds. P. Wester, A. Mishra, A. Mukherji, and A.B. Shrestha, 209–255. Cham: Springer International Publishing.

Bookhagen, B., R.C. Thiede, and M.R. Strecker. 2005. Late Quaternary intensified monsoon phases control landscape evolution in the northwest Himalaya. *Geology* 33 (2): 149–152.

Brahmbhatt, R.M., I.M. Bahuguna, B.P. Rathore, A.V. Kulkarni, R.D. Shah, A.S. Rajawat, and J.S. Kargel. 2017. Significance of glacio-morphological factors in glacier retreat: A case study of part of Chenab basin, Himalaya. *Journal of Mountain Science* 14: 128–141.

Bricker, J.D., W. Schwanghart, B.R. Adhikari, S. Moriguchi, V. Roeber, and S. Giri. 2017. Performance of models for flash flood warning and hazard assessment: The 2015 Kali Gandaki landslide dam breach in Nepal. *Mountain Research Development* 37: 5–15.

Buckman, S., J.C. Aitchison, A. Nutman, V. Bennett, W.M. Saktura, J. Walsh, S. Kachovich, and H. Hidaka. 2018. The Spongtang Massif in Ladakh, NW Himalaya: An early Cretaceous record of spontaneous, intra–oceanic subduction initiation in the Neotethys. *Gondwana Research* 63: 226–249.

Burbank, D.W. and M.B. Fort. 1985. Bedrock control on glacial limits: Examples from the Ladakh and Zanskar ranges, north-western Himalaya, India. *Journal of Glaciology* 31, 143–149.

Burchfield, B.C., C. Zhiliang, K.V. Hodges, L. Yuping, L.H. Royden, D. Changrong, and X. Jiene. 1992. The South Tibetan Detachment System, Himalayan orogen: Extension contemporaneous with and parallel to shortening in a collisional mountain belt. *Geological Society of America Special Paper* 269: 1–41.

Buri, P., and F. Pellicciotti. 2018. Aspect controls the survival of ice cliffs on debris covered glaciers. *Proceedings of National Academy of Sciences* 115 (17): 4369–4374.

Byers, A.C., D.R. Rounce, D.H. Shugar, J.M. Lala, E.A. Byers, and D. Regmi. 2019. A rockfall-induced glacial lake outburst flood, Upper Barun valley, Nepal. *Landslides* 16: 533–549.

Byrne, M.E. 2009. *Glacier Monitoring in Ladakh and Zanskar, Northwestern India.* Graduate Student Theses, Dissertations & Professional Papers. 493. Missoula, The University of Montana.

Capron, E., S.O. Rasmussen, T.J. Popp, T. Erhardt, H. Fischer, A. Landais, J.B. Pedro, G. Vettoretti, A. Grinsted, V. Gkinis, B. Vaughn, A. Svensson, B.M. Vinther, and J.W.C. White. 2021. The anatomy of past abrupt warmings recorded in Greenland ice. *Nature Communications* 12 (1): 2106.

Cartagena-Sierra, A., M.A. Berke, R.S. Robinson, B. Marcks, I.S. Castaneda, A. Starr, I.R. Hall, S.R. Hemming, L.J. LeVay, and Expedition 361 Scientific Party. 2021. Latitudinal migrations of the Subtropical Front at the Agulhas Plateau through the mid-Pleistocene transition. *Paleoceanography and Paleoclimatology* 36: e2020PA004084.

Cavalli, M., T. Heckmann and L. Marchi. 2019. Sediment connectivity in proglacial areas. In *Geomorphology of Proglacial Systems: Landform and Sediment Dynamics in Recently Deglaciated Alpine Landscapes, Geography of the Physical Environment,* eds. T. Heckmann, and D. Morche, 271–287. Cham: Springer International Publishing.

Chakraborty, S., S.K. Bhattacharya, P.S. Ranhotra, A. Bhattacharya, and R. Bhushan. 2006. Palaeoclimatic scenario during Holocene around Sangla valley, Kinnaur northwest Himalaya based on multi proxy records. *Current Science* 91 (6): 777–782.

Chand, P., M.C. Sharma, R. Bhambri, C.V. Sangewar, and N. Juyal. 2017. Reconstructing the pattern of the Bara Shigri glacier fluctuation since the end of the Little Ice Age, Chandra valley, north-western Himalaya. *Progress in Physical Geography* 41: 643–675.

Chand, P., S.K. Jain, H.P. Thakur, S. Kumar, and M.P. Sharma. 2020. Recessional pattern and surface elevation change of the Parvati Glacier, North-Western Himalaya (1965–2018) using remote sensing. *Geocarto International* 41 (24): 9360–9392.

Chandra, R., N. Kowser, M.E. Brookfield, M. Satyanarayanan, and D. Stöckli. 2023. Nature of the Shyok (Northern) Suture Zone between India and Asia: Petrology, geochemistry and origin of the Tirit granitoids and associated dykes (Nubra Valley Ladakh Himalaya, NW India). *Geological Magazine* 160 (5): 1–20.

Chen, F., M. Zhang, H. Guo, S. Allen, J.S. Kargel, U.K. Haritashya, and C.S. Watson. 2021. Annual 30 m dataset for glacial lakes in High Mountain Asia from 2008 to 2017. *Earth System Science Data* 13: 741–766.

Chevuturi, A., A.P. Dimri, and R.J. Thayyen. 2018. Climate change over Leh (Ladakh), India. *Theoretical and Applied Climatology* 131 (1–2): 531–545.

Chiang, F., O. Mazdiyasni, and A. AghaKouchak. 2021. Evidence of anthropogenic impacts on global drought frequency, duration, and intensity. *Nature* 12: 2754.

Chmura, H.E., H.M. Kharouba, J. Ashander, S.M. Ehlman, E.B. Rivest, and L.H. Yang. 2019. The mechanisms of phenology: The patterns and processes of phenological shifts. *Ecological Monographs* 89 (1): e01337.

Civel-Mazens, M., X. Crosta, G. Cortese, E. Michel, A. Mazaud, O. Ther, M. Ikehara, and T. Itaki. 2021. Impact of the Agulhas Return Current on the oceanography of the Kerguelen Plateau region, Southern Ocean, over the last 40kyrs. *Quaternary Science Reviews* 251: 106711.

Cook, K.L., C. Andermann, F. Gimbert, B.R. Adhikari, and N. Hovius. 2018. Glacial lake outburst floods as drivers of fluvial erosion in the Himalaya. *Science* 362: 53–57.

Cook, S.J., D.A. Swift, M.P. Kirkbride, P.G. Knight, and R.I. Waller. 2020. The empirical basis for modelling glacial erosion rates. *Nature Communications* 11: 759.

Cullen, H.M., P.B. de Menocal, S. Hemming, G. Hemming, F.H. Brown, T. Guilderson, and F. Sirocko. 2000. Climate change and the collapse of the Akkadian empire: Evidence from the deep sea. *Geology* 28: 379–382.

Dahal, D.R., M. Mehta, C.M. Bhatt, and K. Luirei. 2024. Flash floods and their cascading tumults: An example from Teesta River valley, Eastern Himalaya, Sikkim, India. *Current Science* 126 (10): 1280–1284.

Damm, B. 2006. Late Quaternary glacier advances in the upper catchment area of the Indus river (Ladakh and western Tibet). *Quaternary International* 154–155: 87–89.

Dar, R.A., O. Jaan, K.O. Murtaza, and S.A. Romshoo. 2017. Glacial-geomorphic study of the Thajwas glacier valley, Kashmir Himalayas, India. *Quaternary International* 444: 157–171.

Das, S. 2022. *Mapping glacier dynamics in the Jankar Chhu Watershed, Lahaul Himalaya.* Unpublished Ph.D. Thesis. New Delhi, Jawaharlal Nehru University.

Das, S., and M.C. Sharma. 2019. Glacier changes between 1971 and 2016 in the Jankar Chhu watershed, Lahaul Himalaya , India. *Journal of Glaciology* 65 (249): 13–28.

Das, S., M.K. Murari, M.C. Milap, R. Saini, M.K. Jaiswal, P.V. Kumar, and P. Kumar. 2023. Reconstraining the Quaternary glacial history of Lahaul Himalaya, northern India. *Quaternary Science Reviews* 316 : 108258.

Davison, B.J., A.J. Sole, S.J. Livingstone, T.R. Cowton, and P.W. Nienow. 2019. The influence of hydrology on the dynamics of land terminating sectors of the Greenland ice sheet. *Frontiers in Earth Science* 7: 10.

Debnath, M., M. Sharma, and H. Syiemlieh. 2019. Glacier dynamics in Changme Khangpu basin, Sikkim Himalaya, India, between 1975 and 2016. *Geoscience* 9 (6): 259.

DeCelles, P.G., B. Carrapa, T.P. Ojha, G.E. Gehrels, and D. Collins. 2020. Structural and thermal evolution of the Himalayan thrust belt in midwestern Nepal. *Geological Society of America,* 547.

Dehecq, A., N. Gourmelen, A. Gardner, F. Brun, D. Goldberg, P.W. Nienow, E. Berthier, C. Vincent, P. Wagnon, and E. Trouvé. 2018. Twenty-first century glacier slowdown driven by mass loss in High Mountain Asia. *Nature Geoscience* 12: 22–27.

Delaney, K.B., and S.G. Evans. 2015. The 2000, Yigong landslide (Tibetan Plateau), rockslide-dammed lake and outburst flood: Review, remote sensing analysis, and process modelling. *Geomorphology* 246: 377–393.

Delaney, I. and L.S. Anderson, 2022. Debris cover limits subglacial erosion and promotes till accumulation. *Geophysical Research Letters* 49: e2022GL099049.

Delaney, I., A. Bauder, M. Werder, and D. Farinotti. 2018. Regional and annual variability in subglacial sediment transport by water for two glaciers in the Swiss Alps. *Frontiers in Earth Science* 6: 175.

Demske, D., P.E. Tarasov, B. Wünnemann, and F. Riedel. 2009. Late Glacial and Holocene vegetation, Indian monsoon and westerly circulation in the trans-Himalaya recorded in the lacustrine pollen sequence from Tso Kar, Ladakh, NW India. *Palaeogeography, Palaeoclimatology, Palaeoecology* 279: 172–185.

Deswal, S., M.P. Sharma, R. Saini, P. Chand, S. Prakash, P. Kumar, D.I. Barr, S.U. Latief, P. Dalal, and I.M. Bahuguna. 2023. Reconstruction of post-Little Ice Age glacier recession in the Lahaul Himalaya, north-west India. *Geografiska Annaler: Series A, Physical Geography* 105 (1): 1–26.

Dhakal, S., L. Srivastava, and B. Sharma, 2019. Meeting future energy needs in the Hindu Kush Himalaya. In *The Hindu Kush Himalayan Assessment: Mountains, Climate Change, Sustainability and People,* eds. P. Wester, A. Mishra, A. Mukherji, and A. Shrestha, 168–202. Springer.

Dhanta, R., and Y.S. Negi. 2018. Agricultural growth and crop diversification in western Himalayan state of Himachal Pradesh. *Pacific Business Review International* 11 (4): 114–123.

Dhar, S., A. Kumar, and S.K. Rai. 2020. Spatio-temporal disposition of Chandra basin glaciers from 1980 to 2011, Lahaul and Spiti Himalayan region, Himachal Pradesh India. *International Journal on Emerging Technologies* 11 (2): 1005–1012.

Di Pietro, J.A. 2004. Tectonostratigraphic subdivisions of the Himalaya: A view from the west. *Tectonics* 23: TC5001.

Dikshit, A., and N. Satyam. 2019. Probabilistic rainfall thresholds in Chibo, India: Estimation and validation using monitoring system. *Journal of Mountain Science* 16: 870–883.

Dikshit, A., R. Sarkar, B. Pradhan, S. Segoni, and A.M. Alamri. 2020. Rainfall induced landslide studies in Indian Himalayan region: A critical review. *Applied Sciences* 10: 2466.

Dilek, Y., and H. Furnes. 2009. Structure and geochemistry of Tethyan ophiolites and their petrogenesis in subduction rollback systems. *Lithos* 113: 1–20.

Dilshad, T., D. Mallick, P.B. Udas, C.G. Goodrich, A. Prakash, G. Gorti, S. Bhadwal, M.Z. Anwar, N. Khandekar, S.M.T. Hassan, N. Habib, S.S. Abbasi, M.A. Syeda, and A. Rahman. 2019. Growing social vulnerability in the river basins: Evidence from the Hindu Kush Himalaya (HKH) Region. *Environmental Development* 31: 19–33.

Dimri, A.P. 2021. Decoding the Karakoram anomaly. *Science of the Total Environment* 788: 147864.

Dimri, A.P., A. Chevuturi, D. Niyogi, R.J. Thayyen, K. Ray, S.N. Tripathi, A.K. Pandey, and U.C. Mohanty. 2017. Cloudbursts in Indian Himalayas: A review. *Earth Science Reviews* 168: 1–23.

Dimri, A.P., D. Kumar, A. Choudhary, and P. Maharana. 2018. Future changes over the Himalayas: Mean temperature. *Global and Planetary Changes* 162: 235–251.

Ding, J., L. Cuo, Y. Zhang, and F. Zhu. 2018. Monthly and annual temperature extremes and their changes on the Tibetan Plateau and its surroundings during 1963–2015. *Scientific Reports* 8: 11860.

Ding, L., R.A. Spicer, J. Yang, Q. Xu, F. Cai, S. Li, Q. Lai, H. Wang, T.E.V. Spicer, Y. Yue, A. Shukla, G. Srivastava, M.A. Khan, S. Bera, and R. Mehrotra. 2017. Quantifying the rise of the Himalaya orogen and implications for the South Asian monsoon. *Geology* 45 (3): 215–218.

Dobriyal, M.J.R., and A. Bijalwan. 2017. Forest fire in western Himalayas of India: A Review. *New York Science Journal* 10 (6): 39–46.

Dolker, P. 2018. An overview of transition in traditional agriculture of Ladakh. *Journal of Himalayan Ecology and Sustainable Development* 13: 26–46.

Draganits, E., S. Gier, C.-Ch. Hofmann, C. Janda, B. Bookhagen, and B. Grasemann. 2014. Holocene versus modern catchment erosion rates at 300 MW Baspa II hydroelectric power plant (India, NW Himalaya). *Journal of Asian Earth Sciences* 90: 157–172.

Dubey, J., N.A. Sheikh, V. Srivastava, A. Sharma, P. Morthekai, R. Singh, R.K. Sharma, P. Pandey, and B. Thakur. 2019. Glacial geomorphology and landscape evolution of the Thangu valley, north Sikkim Himalaya, India. *Journal of the Indian Society of Remote Sensing* 47: 821–837.

Dubey, S. and M.K. Goyal. 2020. Glacial lake outburst flood hazard, downstream impact, and risk over the Indian Himalayas. *Water Resources Research* 56: e2019WR026533.

Dutta, S., S.A.I. Mujtaba, H.S. Saini, R. Chunchekar, and P. Kumar. 2018. Geomorphic evolution of glacier-fed Baspa valley, NW Himalaya: Record of Late Quaternary climate change, monsoon dynamics and glacial fluctuations. In *The Himalayan Cryosphere: Past and Present,* eds. N.C. Pant, R. Ravinder, D. Srivastava, and L.G. Thompson,. Special Publications 462. London, The Geological Society of London.

Dyurgerov, M., M.F. Meier, and D.B. Bahr. 2009. A new index of glacier area change: A tool for glacier monitoring. *Journal of Glaciology* 55 (192): 710–716.

EMDAT Report 2023. *Disasters in Numbers*. Belgium, Centre for Research on Epidemiology of Disasters (CRED).

Emmer, A. 2024. Understanding the risk of glacial lake outburst floods in the twenty-first century. *Nature Water* 2: 608–610.

Farinotti, D., W.W. Immerzeel, R.J. de Kok, D.J. Quincey, and A. Dehecq. 2020. Manifestations and mechanisms of the Karakoram glacier anomaly. *Nature Geoscience* 13: 8–16.

Ferreira, D., P. Cessi, H.K. Coxall, A. de Boer, H.A. Dijkstra, S.S. Drijfhout, T. Eldevik, N. Harnik, J.F. McManus, D.P. Marshall, J. Nilsson, F. Roquet, T. Schneider, and R.C. Wills. 2018. Atlantic-Pacific asymmetry in deep water formation. *Annual Review of Earth and Planetary Sciences.* 46: 327–352.

Ficetola, G.F., S. Marta, A. Guerrieri, I. Cantera, A. Bonin, S. Cauvy-Fraunié, R. Ambrosini, M. Caccianiga, F. Anthelme, R.S. Azzoni, P. Almond, P.A. Gazitúa, J.L.C. Lievano, P. Chand, M.P. Sharma, J.J. Clague, J.A.C. Rapre, C. Compostella, R.C. Encarnación, O. Dangles, P. Deline, A. Eger, S. Erokhin, A. Franzetti, L. Gielly, F. Gili, M. Gobbi, S. Hågvar, R. Kaufmann, N. Khedim, R.I. Meneses, M.A. Morales-Martínez, G. Peyre, F. Pittino, A. Proietto, A. Rabatel, K. Sieron, L. Tielidze, N. Urseitova, Y. Yang, V. Zaginaev, A. Zerboni, A. Zimmer, G.A. Diolaiuti, P. Taberlet, J. Poulenard, D. Fontaneto, W. Thuiller, and A. Carteron. 2024. The development of terrestrial ecosystems emerging after glacier retreat. *Nature* 632 (8024): 336–342.

Finkel, R.C., L.A. Owen, P.L. Barnard, and M.W. Caffee. 2003. Beryllium-10 dating of Mount Everest moraines indicates a strong monsoon influence and glacial synchroneity throughout the Himalaya. *Geology* 31 (6): 561–564.

Fischer, M. 2022. *Outburst floods in the Greater Himalayas: From Regional Susceptibility to Local Hazards*. Ph.D. Dissertation, University of Potsdam, Germany.

Fischer, E.M., S. Sippel, and R. Knutti. 2021. Increasing probability of record-shattering climate extremes. *Nature Climate Change* 11: 689–695.

Forster, P.M., H.I. Forster, M.J. Evans, M.J. Gidden, C.D. Jones, C.A. Keller, R.D. Lamboll, C. Le Quéré, J. Rogelj, D. Rosen, C.-F. Schleussner, T.B. Richardson, C.J. Smith, and S.T. Turnock. 2020. Current and future global climate impacts resulting from COVID-19. *Nature Climate Change* 10: 913–919.

Fort, M. 1983. Geomorphological observations in the Ladakh arc (Himalayas): Quaternary evolution and present dynamics. In *Stratigraphy and Structure of the Kashmir and Ladakh, Himalaya*, ed. V.J. Gupta, 39–58. New Delhi: Hindustan Publishing Corporation.

Forte, A.M., J.S. Leonard, M.W. Rossi, K.X. Whipple, A.M. Heimsath, L. Sukhishvil, T. Godoladze, and F. Kadirov. 2022. Low variability runoff inhibits coupling of climate, tectonics, and topography in the Greater Caucasus. *Earth and Planetary Science Letter* 584: 117525.

Fowler, A.C. 2000. An instability mechanism for drumlin formation. In *Deformation of Glacial Materials,* eds. A.J. Maltman, B. Hubbard, and M.T. Hambrey, 307–319. Geological Society, London, Special Publication 176.

Fowler, H.J. and D.R. Archer. 2005. Hydro-climatological variability in the Upper Indus Basin and implication for water resources. In *Regional Hydrological Impacts of Climatic Change-Impact Assessment and Decision Making,* eds. T. Wagener, S. Franks, H.V. Gupta, E. Bøgh, L. Bastidas, C. Nobre, C.de O. Galvão, 131–138. International Association of Hydrological Science Publication 295.

Fox-Kemper, B., H.T. Hewitt, C. Xiao, G. Aðalgeirsdóttir, S.S. Drijfhout, T.L. Edwards, N.R. Golledge, M. Hemer, R.E. Kopp, G. Krinner, A. Mix, D. Notz, S. Nowicki, I.S. Nurhati, L. Ruiz, J.-B. Sallée, A. B.A. Slangen, and Y. Yu. 2021. Ocean, cryosphere and sea level change. In *Climate Change 2021: The Physical Science basis. Contribution of Working Group I to the Sixth Assessment Report of the Intergovernmental Panel on Climate Change*, eds. V. Masson-Delmotte, P. Zhai, A. Pirani, S.L. Connors, C. Péan, S. Berger, N. Caud, Y. Chen, L. Goldfarb, M.I. Gomis, M., Huang, K., Leitzell, E., Lonnoy, J.B.R., Matthews, T.K. Maycock, T. Waterfield, O. Yelekçi, R. Yu, and B. Zhou, 1211–1362. Cambridge University Press, Cambridge, United Kingdom and New York, NY, USA.

Froude, M.J., and D.N. Petley. 2018. Global fatal landslide occurrence from 2004 to 2016. *Natural Hazards Earth System Science* 18: 2161–2181.

FSI (Forest Survey of India). 2019. *India State of Forest Report 2019.* Dehradun: Forest Survey of India, Ministry of Environment, Forests and Climate Change.

Fujita, K., H. Inoue, T. Izumi, S. Yamaguchi, A. Sadakane, S. Sunako, K. Nishimura, W.W. Immerzeel, J.M. Shea, R.B. Kayastha, T. Sawagaki, D.F. Breashears, H. Yagi, and A. Sakai. 2017. Anomalous winter-snow-amplified earthquake induced disaster of the 2015 Langtang avalanche in Nepal. *Natural Hazards and Earth System Sciences* 17: 749–764.

Fuller, S. and T. Murray. 2020. Evidence against pervasive bed deformation during the surge of an Icelandic glacier. In *Deformation of Glacial Materials,* eds. A.J. Maltman, B. Hubbard, and M.T. Hambrey, 203–216. Geological Society, London, Special Publications 176.

Ganjoo, R.K. 2009. Holocene tectonics and climate of Durung Drung glacier basin, Zanskar Himalaya, India (Abstract). The 5th International Symposium on Tibetan Plateau and 24th Himalaya-Karakoram-Tibet Workshop, 11–14 Aug. Beijing, China.

Ganjoo, R.K. 2015. Early human adaptation vis-à-vis high Himalayan landscape and climate, Ladakh Himalaya, Jammu & Kashmir. *Journal of Indian Geomorphology* 3: 5–9.

Ganjoo, R.K. 2021. Glacier fed Himalayan rivers of India. *Current Science* 121 (2): 210–213.

Ganjoo, R.K., and M.N. Koul. 2005. Glacial meltwater impounding: Evidence from the late Quaternary glaciogenic sediments in the Sangla valley, district Kinnaur, Himachal Pradesh, India. *Journal of Earth System Science* 114 (4): 381–385.

Ganjoo, R.K., and M.N. Koul. 2009. Is the Siachen glacier melting? *Current Science* 97 (3): 309–310.

Ganjoo, R.K. and M.N. Koul. 2013. Asynchronous behavior of glaciers of Ladakh Himalaya, J & K State, India. In *Earth System Processes and Disaster Management Society of Earth Scientists Series 1* eds. R. Sinha, and R. Ravindra, 25-36. Berlin, Springer-Verlag.

Ganjoo, R.K., M.N. Koul, I.M. Bahuguna, and Ajai (2014). The complex phenomenon of glaciers of Nubra valley, Karakoram (Ladakh), India. *Natural Science* 6: 733–740.

Ganju, A., Y.C. Nagar, L.N. Sharma, and S. Sharma. 2018. Luminescence chronology and climatic implication of the late quaternary glaciation in the Nubra valley, Karakoram Himalaya, India. *Palaeogeography, Palaeoclimatology, Palaeoecology* 502 (6): 52–62.

Gao, H., X. Zou, J. Wu, Y. Zhang, X. Deng, S. Hussain, M.A. Wazir, and G. Zhu. 2020. Post-20th century near-steady state of Batura glacier: Observational evidence of Karakoram anomaly. *Scientific Reports* 10: 987.

Gao, Y., S. Zhao, D. Jha, and M.M. Rahman. 2021. Flood assessment and early warning of the reoccurrence of river blockage at the Baige landslide. *Journal of Geographical Sciences* 31 (11): 1694–1712.

García-Hernández, C., J. Ruiz-Fernández, C. Sánchez-Posada, S. Pereira, M. Oliva, and G. Vieira. 2017. Reforestation and land use change as drivers for a decrease of avalanche damage in mid-latitude mountains (NW Spain). *Global and Planetary Change* 153: 35–50.

Garg, P.K., A. Shukla, R.K. Tewari, and A.S. Jasrotia. 2017. Assessing the status of glaciers in parts of the Chandra basin, Himachal Himalaya: A multiparametric approach. *Geomorphology* 284: 99–114.

Garg, S., A. Shukla, M. Mehta, V. Kumar, S.A. Samuel, S.K. Bartarya, and U.K. Shukla. 2018. Field evidences showing rapid frontal degeneration of the Kangriz glacier, western Himalayas, Jammu & Kashmir. *Journal of Mountain Science* 15: 1199–1208.

Garg, S., A. Shukla, M. Mehta, V. Kumar, and U.K. Shukla. 2019. On geomorphic manifestations and glaciation history of the Kangriz glacier, western Himalaya. *Himalayan Geology* 40 (2): 115–127.

Garg, S., A. Shukla, P.K. Garg, B. Yousuf, and U.K. Shukla. 2022a. Surface evolution and dynamics of the Kangriz glacier, western Himalaya in past 50 years. *Cold Regions Science and Technology* 196: 103496.

Garg, P.K., J.S. Yadav, S.K. Rai, and A. Shukla. 2022b. Mass balance and morphological evolution of the Dokriani Glacier, central Himalaya, India during 1999–2014. *Geoscience Frontiers* 13 (1): 101290.

Gebregiorgis, D., E.C. Hathorne, L. Giosan, S. Clemens, D. Nürnberg, and M. Frank. 2018. Southern Hemisphere forcing of South Asian monsoon precipitation over the past ~1 million years. *Nature Communications* 9: 4702.

Ghosh, B., D. Bandyopadhyay, and T. Morishita. 2017. Andaman-Nicobar ophiolites, India: Origin, evolution and emplacement. *Geological Society London Memoir* 47: 95–110.

Gioli, G., G. Thapa, F. Khan, P. Dasgupta, D. Nathan, L. Adhikari, N. Chhetri, S.K. Mohanty, E. Aurino, and L.M. Scott. 2018. Understanding and tackling poverty and vulnerability in mountain livelihood in the Hindu Kush Himalaya. In *The Hindu Kush Himalayan Assessment: Mountains, Climate change, Sustainability and People,* eds. P. Wester, A. Mishra, A. Mukherji and A. Shrestha, 421–455. Springer.

Graham, D.J. and N.G. Midgley. 2000. Moraine-mound formation by englacial thrusting: the Younger Dryas moraines of Cwm Idwal, North Wales. In *Deformation of Glacial Materials,* eds. A.J. Maltman, B. Hubbard, and M.T. Hambrey, 322–336. Geological Society, London, Special Publications 176.

Graven, H., R.F. Keeling, and J. Rogelj, 2020. Changes to carbon isotopes in atmospheric $CO_2$ over the industrial era and into the future. *Global Biogeochemical Cycles* 34(11): e2019GB006170.

Guhathakurta, P., B.L. Kumar, P. Menon, A.K. Prasad, S.T. Sable, and S.C. Advani, 2020. *Observed Rainfall Variability and Changes over Uttarakhand State.* Pune, Climate Research and Services, Indian Meteorology Department.

Gujree, I., I. Ahmad, F. Zhang, and A. Arshad. 2022. Innovative trend analysis of high-altitude climatology of Kashmir valley North-West Himalayas. *Atmosphere* 13: 764.

Gupta, S., A. Roy, D. Bhavsar, R. Kala, S. Singh, and A.S. Kumar. 2018. Forest fire burnt area assessment in the biodiversity rich regions using geospatial technology: Uttarakhand forest fire event 2016. *Journal of the Indian Society of Remote Sensing* 46: 945–955.

Hager, C. 2003. *Out-of-sequence Extrusion in the Sutlej Valley (NW-Himalaya).* Diploma Thesis. Vienna, University of Vienna.

Hakhoo, N., G.M. Bhat, S. Pandita, G. Hussain, A. Ul Haq, M. Hafiz, W. Ahmed, Y. Singh, and B. Thusu. 2019. Natural hazards—Their drivers, mechanisms and impacts in the Shyok-Nubra valley, NW Himalaya, India. *International Journal of Disaster Risk Reduction* 35: 101094.

Hall, M.L., A.C.K. Lee, C. Cartwright, S. Marahatta, J. Karki, and P. Simkhada. 2017. The 2015 Nepal earthquake disaster: Lessons learned one year on. *Public Health* 145: 39–44.

Hanusch, M., X. He, V. Ruiz-Hernández, and R.R. Junker. 2022. Succession comprises a sequence of threshold-induced community assembly processes towards multidiversity. *Nature Communications Biology* 5: 424.

Haque, D.M.E., T. Hayat, and S. Tasnim. 2019. Time series analysis of subsidence in Dhaka city, Bangladesh using INSAR. *Malaysian Journal of Geosciences* 3 (1): 32–44.

Hayat, H., B. Raup, S. Muhammad, S. Liu, R. Khan, S.U. Baig, and A.A. Tahir. 2023. Heterogeneous mass balance of selected glaciers in the Hindu Kush, Karakoram, and Himalaya between 2000 and 2018. *European Journal of Remote Sensing* 56 (1): 2173086.

Herreid, S. 2021. What can thermal imagery tell us about glacier melt below rock debris? *Frontiers in Earth Sciences* 9: 681059.

Herreid, S., and F. Pellicciotti. 2020. The state of rock debris covering earth's glaciers. *Nature Geoscience* 13 (9): 621–627.

Hewitt, K. 2011. Glacier change, concentration, and elevation effects in the Karakoram Himalaya, Upper Indus Basin. *Mountain Research and Development* 31 (3): 188–200.

van der Horst, T., M.M. Rutten, N. van de Giesen, and R.F. Hanssen. 2018. Monitoring land subsidence in Yangon, Myanmar using Sentinel-1 persistent scatterer interferometry and assessment of driving mechanisms. *Remote Sensing of Environment* 217: 101–110.

Huang, R., H. Zhu, E. Liang, A. Bräuning, L. Zhong, C. Xu, X. Feng, F. Asad, S.R. Sigdel, L. Li, and J. Grießinger. 2022. Contribution of winter precipitation to tree growth persists until the late growing season in the Karakoram of northern Pakistan. *Journal of Hydrology* 607: 127513.

Hunt, K.M.R., A.G. Turner, and L.C. Shaffrey. 2018. The evolution, seasonality and impacts of western disturbances . *Quarterly Journal of the Royal Meteorological Society* 144 (710): 278–290.

Hunt, K.M.R., A.G. Turner, and R.K.H. Schiemann. 2021. How interactions between tropical depressions and western disturbances affect heavy precipitation in South Asia. *Monthly Weather Review* 149 (6): 1801–1825.

Hunt, K.M.R., J.-P. Baudouin, A.G. Turner, A.P. Dimri, G. Jeelani, Pooja, R. Chattopadhyay, F. Cannon, T. Arulalan, M.S. Shekhar, T.P. Sabin and E. Palazzi. 2024. (preprint). Western disturbances and climate variability: A review of recent developments, *EGUsphere.*

Huo, D., M.P. Bishop, and A.B.G. Bush. 2021. Understanding complex debris-covered glaciers: Concepts, issues, and research directions. *Frontiers in Earth Science* 9: 652279.

Hussain, A., and F.M. Qamar. 2020. Dual challenge of climate change and agrobiodiversity loss in mountain food systems in the Hindu Kush Himalaya. *One Earth* 3: 539–542.

Immerzeel, W.W., L.P.H. van Beek, and M.F.P. Bierkens. 2010. Climate change will affect the Asian water towers. *Science* 328 (5984): 1382–1386.

Indian Meteorological Department. 2014. *Climate of Jammu and Kashmir*. Pune, Indian Meteorological Department.

IPCC. 2019. *Special Report on the Ocean and Cryosphere in a Changing Climate.* Intergovernmental Panel on Climate Change.

ISFR. 2023. *Indian State of Forest Report Volume II*. Dehradun, Forest Survey of India.

Islam, N., and P.P. Patel. 2022. Inventory and GLOF hazard assessment of glacial lakes in the Sikkim Himalayas, India. *Geocarto International* 37 (13): 3840–3876.

Jain, A.K., D. Kumar, S. Singh, A. Kumar, and N. Lal. 2000. Timing, quantification and tectonic modelling of Pliocene-Quaternary movements in the NW Himalaya: Evidence from fission track dating. *Earth and Planetary Science Letters* 179: 437–451.

Jain, A.K., T. Ahmad, S. Singh, S.K. Ghosh, R.C. Patel, R. Kumar, K.K. Agarwal, J. Perumal, R. Islam, and O.N. Bhargava. 2012. Evolution of the Himalaya. *Proceedings of Indian National Science Academy* 78 (3): 259–275.

Javed, A., K. Anshuman, P. Kumar, and D. Sachan. 2023. The decline in western disturbance activity over Northern India in recent decades. *Climate Change* 176: 97.

Jagoutz, O., P. Bouilhol, U. Schaltegger, and O. Muntener. 2018. The isotopic evolution of the Kohistan Ladakh arc from subduction initiation to continent arc collision. *Geological Society of London* Special Publication 483, SP483–7.

Jasrotia, A.S., S. Ahmad, P.K. Thakur, Q. Ridwan, Z.A. Wani, S.A.M. Alamri, S. Siddiqui, and M. Moustafa. 2023. Long-term geospatial observations of the Drang Drung and Pensilungpa glaciers, north western Himalaya, India, from 1976 to 2020. *Sustainability* 15: 15067.

Jepson, G., B. Carrapa, J. Gillespie, R. Feng, P.G. DeCelles, P. Kapp, C.R. Tabor, and J. Zhu. 2021. Climate as the great equalizer of continental-scale erosion. *Geophysical Research Letters* 48: e2021GL095008.

Ji, J.F., J. Shen, W. Balsam, J. Chen, L. Liu, and X. Liu. 2005. Asian monsoon oscillations in the northeastern Qinghai-Tibet Plateau since the late glacial as interpreted from visible reflectance of Qinghai lake sediments. *Earth and Planetary Science Letters* 233 (1–2): 61–70.

Jiskoot, H., C.J. Curran, D.L. Tessler, and L.R. Shenton. 2009. Changes in Clemenceau icefield and Chaba group glaciers, Canada, related to hypsometry, tributary detachment, length–slope and area–aspect relations. *Annals of Glaciology* 50: 133–143.

Joshi, M., and V.C. Thakur. 2016. Signatures of 1905 Kangra and 1555 Kashmir earthquakes in Medieval period temples of Chamba region, northwest Himalaya. *Seismological Research Letters* 87 (5): 1–11.

Joshi, S., and S.S. Subramanian, 2023. Earthquake-induced landslides in the Himalayas: A review of selected cases. *Conference Paper at 3rd International Conference on Earthquake Engineering and Post Disaster Reconstruction Planning (ICEE-PDRP 2023)*, October 2023, Bhaktapur, Nepal.

Joshi, L.M., B.S. Kotlia, G.C. Kothyari, A.K. Singh, A.K. Taloor, R. Upadhyaya, and D. Dayal. 2021. Neotectonic landform development and associated mass movements along eastern Ramganga valley in the Kumaun Himalaya, India. *Geotectonics* 55 (4): 543–562.

Kääb, A., M. Jacquemart, A. Gilbert, S. Leinss, L. Girod, C. Huggel, D. Falaschi, F. Ugalde, D. Petrakov, S. Chernomorets, M. Dokukin, F. Paul, S. Gascoin, E. Berthier, and J. Kargel. 2021. Sudden large-volume detachments of low-angle mountain glaciers—More frequent than thought. *The Cryosphere* 15: 1751–1785.

Kanth, T.A., A.A. Shah, and Z. Hassan. 2011. Geomorphologic character & receding trend of Kolahai glacier in Kashmir Himalaya. *Recent Research in Science and Technology* 3 (9): 68–73.

Karki, R., S. Hasson, L. Gerlitz, R. Talchabhadel, U. Schickhoff, T. Scholten, and J. Böhner. 2020. Rising mean and extreme near-surface air temperature across Nepal. *International Journal of Climatology* 40: 2445–2463.

Karmakar, N., A. Chakraborty, and R. Nanjundiah. 2017. Increased sporadic extremes decrease the intra-seasonal variability in the Indian summer monsoon rainfall. *Scientific Reports* 7: 7824.

Kathayat, G., H. Cheng, A. Sinha, L. Yi, X. Li, H. Zhang, H. Li, Y. Nong, and R.L. Edwards. 2017. The Indian monsoon variability and civilization changes in the Indian subcontinent. *Science Advances* 3 (12): e1701296.

Kaul, M.N. 1990. *Glacial and Fluvial Geomorphology of Western Himalayas*. New Delhi: Concept Publishing Company.

Kaveh-Firouz, A., J.-P. Burg, N. Haghipour, S.K. Mandal, M. Christl, and A. Mohammadi. 2023. Tectonics, base-level fluctuations, and climate impact on the Eocene to present-day erosional pattern of the Arabia-Eurasia collision zone (NNW Iranian Plateau and west Alborz Mountains). *Tectonics* 42: e2022TC007684.

Kearsey, T.I., J.R. Lee, A. Finlayson, M. Garcia-Bajo, and A.A.M. Irving. 2018. Examining the geometry, age and genesis of buried Quaternary valley systems in the Midland Valley of Scotland, UK. *Boreas* 48: 658–677.

Kellett, D.A., J.M. Cottle, and K.P. Larson, 2018. The South Tibetan detachment system: History, advances, definition and future directions. In *Himalayan Tectonics: A Modern Synthesis* eds. Treloar, P.J. and M.P. Searle, 377–400. London: Geological Society, London, Special Publication 483.

Khadka, N., X. Chen, S. Sharma, and B. Shrestha. 2023. Climate change and its impacts on glaciers and glacial lakes in Nepal Himalayas. *Regional Environmental Change* 23: 143.

Khali, H., P. Pandey, S.N. Ali, and P.K. Champatiray. 2022. MODIS observed spatiotemporal variation of snow cover in Zanskar valley, North-Western Himalaya. *Frontiers in Water* 4: 853146.

Khalid, M.A. 2022. Developmental success and the sustainability challenge in a mountain region: Case of Himachal Pradesh in India. *International Journal of Regional Development* 9 (1): 47–74.

Khanal, S., A.F. Lutz, P.D.A. Kraaijenbrink, B. van den Hurk, T. Yao, and W.W. Immerzeel. 2021. Variable 21st century climate change response for rivers in High Mountain Asia at seasonal to decadal time scales. *Water Resources Research* 57: e2020WR029266.

King, O., A. Bhattacharya, R. Bhambri, and T. Bolch. 2019. Glacial lakes exacerbate Himalayan glacier mass loss. *Scientific Reports* 9 (1): 1–9.

Koul, M.N., and R.K. Ganjoo. 2009. Quaternary glacial history of Baspa valley, Kinnar Himalaya, Himachal Pradesh. In *Geomorphology in Inda, Prof. Savindra Singh Felicitation Volume,* eds. H.S. Sharma, and V.S. Kale, 425–440. Allahabad, Prayag Pustak Bhawan.

Koul, M.N., and R.K. Ganjoo. 2010. Impact of inter-and intra-annual variation in weather parameters on mass balance and equilibrium line of Naradu glacier (Himachal Pradesh) NW Himalaya, India. *Climate Change* 99: 119–139.

Koul, M.N., I.M. Bahuguna, Aja, A.S. Rajawat, S. Ali and Koul, S. 2016. Glacier area change over past 50 years to stable phase in Drass valley, Ladakh Himalaya (India). *American Journal of Climate Change* 5(1), 88–102.

Kraaijenbrink, P.D.A., M.F.P. Bierkens, A.F. Lutz, and W.W. Immerzeel. 2017. Impact of a global temperature rise of 1.5°C on Asia's glaciers. *Nature* 549: 257–260.

Krishna Bahadur, K.C., E.D.G. Fraser, D. Wassmandorf, and E. Tzadok. 2022. Climate change creates opportunities to expand agriculture in the Hindu Kush Himalaya but will cause considerable ecosystem trade-offs. *Environmental Research Communications* 4: 111001.

Krishnan, R., A.B. Shrestha, G. Ren, R. Rajbhandari, S. Saeed, J. Sanjay, M.A. Syed, R. Vellore, Y. Xu, Q. You, and Y. Ren. 2019a. Unravelling climate change in the Hindu Kush Himalaya: Rapid warming in the mountains and increasing extremes. In *The Hindu Kush Himalaya Assessment: Mountains, Climate Change, Sustainability and People,* eds. P. Wester, A. Mishra, A. Mukherji, and A. Shrestha, 57–98. Springer.

Krishnan, R., T.P. Sabin, R.K. Madhura, R.K. Vellore, M. Mujumdar, J. Sanjay, S. Nayak, and M. Rajeevan. 2019b. Non-monsoonal precipitation response over the western Himalayas to climate change. *Climate Dynamics* 52: 4091–4109.

Kulkarni, A.V., M. Thamban, T.S. Shirsat, A. Kulkarni, H.S. Negi, and I.M. Bahuguna. 2021. State of Himalayan cryosphere and implications for water security. *Water Security* 14: 100101.

Kumar, P., and K. Krishna. 2021. Temporal fluctuations and frontal changes of Miyar, Takdung and Uldhampu glaciers, Miyar valley, Himachal Pradesh, India. *Proceedings Indian National Science Academy* 87: 1–7.

Kumar, S., and A. Kumar. 2022. Hotspot and trend analysis of forest fires and its relation to climatic factors in the western Himalayas. *Natural Hazards* 114: 3529–3544.

Kumar, R., S.K. Ghosh, and S.J. Sangode. 2011. Sedimentary architecture of late Cenozoic Himalayan foreland basin fill: An overview. *Memoirs of the Geological Society of India* 78: 245–280.

Kumar, S., P.K. Srivastava, and Snehmani. 2017. GIS-based MCDA–AHP modelling for avalanche susceptibility mapping of Nubra valley region, Indian Himalaya. *Geocarto International* 32: 1254–1267.

Kumar, V., V. Gupta, and I. Jamir. 2018. Hazard evaluation of progressive Pawari landslide zone, Satluj valley, Himachal Pradesh, India. *Natural Hazards* 93: 1029–1047.

Kumar, A., H.S. Negi, and K. Kumar. 2020. Long-term mass balance modeling (1986–2018) and climate sensitivity of Siachen glacier, east-Karakoram. *Environmental Monitoring and Assessment* 192: 1–16.

Kumar, V., M. Mehta, A. Shukla, A. Kumar, and S. Garg. 2021a. Late Quaternary glacial advances and equilibrium-line altitude changes in a semi-arid region, Suru Basin, western Himalaya. *Quaternary Science Reviews* 267: 107100.

Kumar, D., A.K. Singh, and D.S. Singh. 2021b. Spatio-temporal fluctuations over Chorabari glacier, Garhwal Himalaya, India between 1976 and 2017. *Quaternary International* 575–576: 178–189.

Kumar, V., I. Jamir, V. Gupta, and R.K. Bhasin. 2021c. Inferring potential landslide damming using slope stability, geomorphic constraints, and run-out analysis: A case study from the NW Himalaya. *Earth Surface Dynamics* 9 (2): 351–377.

Lai, J., and A.M. Anders. 2020. Tectonic controls on rates and spatial patterns of glacial erosion through geothermal heat flux. *Earth and Planetary Science Letters* 543: 116348.

Lai, J., and A.M. Anders. 2021. Climatic controls on mountain glacier basal thermal regimes dictate spatial patterns of glacial erosion. *Earth Surface Dynamics* 9: 845–859.

Latorre, C., J. Wilmshurst, and L. von Gunten (Eds.) 2016. Climate change and cultural evolution. *Past Global Changes* 24(2): 1–83.

Lee, S.Y., Y.B. Seong, L.A. Owen, M.K. Murari, H.S. Lim, H.I. Yoon, and K.C. Yoo. 2014. Late Quaternary glaciation in the Nun-Kun massif, northwestern India. *Boreas* 43 (1): 67–89.

Lee, E., J.L. Carrivick, D.J. Quincey, S.J. Cook, W.H.M. James, and L.E. Brown. 2021. Accelerated mass loss of Himalayan glaciers since the Little Ice Age. *Scientific Reports* 11: 24284.

Leech, M.L., S. Singh, A.K. Jain, S.L. Klemperer, and R.M. Manickavasagam. 2005. The onset of India-Asia continental collision: Early, steep subduction required by the timing of UHP metamorphism in the western Himalaya. *Earth and Planetary Science Letters* 234 (1–2): 83–97.

Lei, Y., K. Yang, B. Wang, Y. Sheng, B.W. Bird, G. Zhang, and L. Tian. 2014. Response of inland lake dynamics over the Tibetan Plateau to climate change. *Climatic Change* 125 (2): 281–290.

Lewington, E.L.M., S.J. Livingstone, C.D. Clark, A.J. Sole, and R.D. Storrer. 2020. A model for interaction between conduits and surrounding hydraulically connected distributed drainage based on geomorphological evidence from Keewatin, Canada. *The Cryosphere* 14: 2949–2976.

Lhakpa, D., Y. Fan, and Y. Cai. 2022. Continuous Karakoram glacier anomaly and its response to climate change during 2000–2021. *Remote Sensing* 14: 6281.

Li, Y., Y.-F. Cui, J.-S. Hao, Z.-T. Zhang, H. Wang, J. Guo, and S.-F. Wang. 2024. Frequency and size change of ice-snow avalanches in the central Himalaya: A case from the Annapurna II glacier. *Advances in Climate Change Research* 15: 464–475.

Lin, H., G. Li, L. Cuo, A. Hooper, and Q. Ye. 2017. A decreasing glacier mass balance gradient from the edge of the upper Tarim basin to the Karakoram during 2000–2014. *Scientific Reports* 7: 6712.

Liu, Z.H., A.C.G. Henderson, and Y.S. Huang. 2006. Alkenone-based reconstruction of late Holocene surface temperature and salinity changes in lake Qinghai China. *Geophysical Research Letters* 33 (9): L09707.

Liu, Z.H., A.C.G. Henderson, and Y.S. Huang. 2008. Regional moisture source changes inferred from late-Holocene stable isotope records. *Advances in Atmospheric Sciences* 25 (6): 1021–1028.

Liu, X., J. Vandenberghe, Z. An, Y. Li, Z. Jin, J. Dong, and Y. Sun. 2016. Grain size of lake Qinghai sediments: Implications for riverine input and Holocene monsoon variability. *Palaeogeography, Palaeoclimatology, Palaeoecology* 449: 41–51.

Liu, Y., M. Lu, H. Yang, A. Duan, B. He, S. Yang, and G. Wu. 2020. Land-atmosphere-ocean coupling associated with the Tibetan Plateau and its climate impacts. *National Science Review* 7: 534–552.

Liu, D., Y. Cui, H. Wang, W. Jin, C. Wu, N.A. Bazai, G. Zhang, P.A. Carling, and H. Chen. 2021a. Assessment of local outburst flood risk from successive landslides: Case study of Baige landslide-dammed lake, upper Jinsha river, eastern Tibet. *Journal of Hydrology* 599: 126294.

Liu, J., M. Lecoq, and L. Zhang. 2021b. Desert locust stopped by Tibetan Highlands during the 2020 upsurge. *Agronomy* 11: 2287.

Lone, J.M., P. Masilamani, and M.A.Q. Khan. 2018. Evaluation of automated and manual methods for measuring change in Kolahai glacier using geospatial technology. *Asian Journal of Science and Technology* 09 (07): 8475–8484.

Lone, S.A., G. Jeelani, V. Padhya, and R.D. Deshpande. 2022. Identifying and estimating the sources of river flow in the cold arid desert environment of Upper Indus River Basin (UIRB), western Himalayas. *Science of the Total Environment* 832: 154964.

Lu, W., X. Zhao, X. Feng, N. Xiang, Z. Du, and W. Zhang. 2022. Temporal and spatial response of Holocene temperature to solar activity. *Quaternary International* 613: 39–45.

Lutz, A.F., W.W. Immerzeel, A.B. Shrestha, and M.F.P. Bierkens. 2014. Consistent increase in High Asia's runoff due to increasing glacier melt and precipitation. *Nature Climate Change* 4: 587–592.

Lutz, A.F., W.W. Immerzeel, P.D.A. Kraaijenbrink, A.B. Shrestha, and M.F.P. Bierkens. 2016. Climate change impacts on the upper Indus hydrology: Sources, shifts and extremes. *PLoS One* 11 (11): e0165630.

MacManus, K., D. Balk, H. Engin, G. McGranahan, and R. Inman. 2021. Estimating population and urban areas at risk of coastal hazards, 1990–2015: How data choices matter. *Earth System Science Data* 13: 5747–580.

Maharjan, S.B., J.F. Steiner, A.B. Shrestha, A. Maharjan, S. Nepal, M. S. Shrestha, B. Bajracharya, G. Rasul, M. Shrestha, M. Jackson, and N. Gupta. 2021. *The Melamchi Flood Disaster: Cascading Hazard and the Need for Multihazard Risk Management.* International Centre for Integrated Mountain Development (ICIMOD), Kathmandu.

Malik, I.H. 2022. Spatial dimension of impact, relief, and rescue of the 2014 flood in Kashmir Valley. *Natural Hazards* 110 (3): 1911–1929.

Malik, J.N., S.P. Naik, S. Sahoo, K. Okumura, and A. Mohanty. 2017. Paleoseismic evidence of the CE 1505 (?) and CE 1803 earthquakes from the foothill zone of the Kumaon Himalaya along the Himalayan Frontal Thrust (HFT), India. *Tectonophysics* 714–715: 133–145.

Malik, I.H., R. Ahmed, J.D. Ford, M.S.A. Shakoor, and S.N. Wani. 2024. Beyond the banks and deluge: Understanding riverscape, flood vulnerability, and responses in Kashmir. *Natural Hazards* 9. https://doi.org/10.1007/s11069-024-06712-z

Mandal, A., B.D. Vishwakarma, T. Angchuk, M.F. Azam, P.K. Garg, and M. Soheb. 2024. Glacier mass balance and its climatic and nonclimatic drivers in the Ladakh region during 2000–2021 from remote sensing data. *Journal of Glaciology* 1–23.

Manish, K., and M.K. Pandit. 2018. Geophysical upheavals and evolutionary diversification of plant species in the Himalaya. *PeerJ* 6: e5919.

Marazi, A., and S.A. Romshoo. 2018. Stremflow response to shrinking glaciers under changing climate in Lidder valley, Kashmir Himalayas. *Journal of Mountain Science* 15: 1241–1253.

Mariotti, A., P.-H. Blard, J. Charreau, S. Toucanne, S.J. Jorry, S. Molliex, D.L. Bourlès, G. Aumaître, and K. Keddadouche. 2021. Nonlinear forcing of climate on mountain denudation during glaciations. *Nature Geosciences* 14, 16–22.

Matin, M., V. Chitale, S.R. Murthy, M. Uddin, K. Bajracharya, Pradhan, and B. Sudip. 2017. Understanding forest fire patterns and risk in Nepal using remote sensing, geographic information system and historical fire data. *International Journal of Wildland Fire* 26(4): 276.

Mayewski, P.A., L.B. Perry, T. Mathews, and S.D. Birkel. 2020. Climate change in the Hindu Kush Himalayas: Basis and gaps. *One Earth* 3: 551–555.

Mehta, M., V. Kumar, S. Garg, and A. Shukla. 2021. Little ice age glacier extent and temporal changes in annual mass balance (2016–2019) of Pensilungpa glacier, Zanskar Himalaya. *Regional Environmental Change* 21 (2): 38.

Menzel, P., B. Gaye, P.K. Mishra, A. Anoop, N. Basavaiah, N. Marwan, B. Plessen, S. Prasad, N. Riedel, M. Stebich, and M.G. Wiesnera. 2014. Linking Holocene drying trends from Lonar lake in monsoonal central India to North Atlantic cooling events. *Palaeogeography, Palaeoclimatology, Palaeoecology* 410: 164–178.

Menzies, J. 2000. Micromorphological analyses of microfabrics and microstructures indicative of deformation processes in glacial sediments. In *Deformation of Glacial Materials,* eds. A.J. Maltman, B. Hubbard, and M.T. Hambrey, 245–257. Geological Society, London, Special Publication 176.

Mina, U., A.P. Dimri, and S. Farswan. 2023. Forest fires and climate attributes interact in central Himalayas: An overview and assessment. *Fire Ecology* 19: 14.

Ministry of Science, and Technology and Environment. 2015. *Nepal Earthquake 2015: Rapid Environmental Assessment*. Kathmandu: Government of Nepal.

Mir, R.A. 2018. Recent changes of two parts of Kolahai glacier and its controlling factors in Kashmir basin, western Himalaya. *Remote Sensing Applications: Society and Environment* 11: 265–281.

Mir, R.A., and Z. Majeed. 2018. Frontal recession of Parkachik glacier between 1971–2015, Zanskar Himalaya using remote sensing and field data. *Geocarto International* 33 (2): 163–177.

Mir, R.A., S.K. Jain, A.K. Lohani, and A. Saraf. 2018a. Glacier recession and glacial lake outburst flood studies in Zanskar basin, western Himalaya. *Journal of Hydrology* 564 (1): 376–396.

Mir, R.A., S.K. Jain, S. Jain, K.T. Renoj, and A. Saraf. 2018b. Assessment of recent glacier changes and its controlling factors from 1976 to 2011 in Baspa basin, Western Himalaya. *Arctic, Antarctic and Alpine Research* 49 (4): 621–647.

Mishra, P.K., A. Anoop, G. Schettler, S. Prasad, A. Jehangir, P. Menzel, R. Naumann, A.R. Yousuf, N. Basavaiah, K. Deenadayalan, M.G. Wiesner, and B. Gaye. 2015. Reconstructed late quaternary hydrological changes from lake Tso Morari, NW Himalaya. *Quaternary International* 371: 76–86.

Modgil, R., S. Kaundal, and A. Sandal. 2019. Bio-chemical and functional characteristics of black gram (*Vigna mungo*) cultivars grown in Himachal Pradesh, India. *International Journal of Current Microbiology and Applied Sciences* 8 (4): 2126–2137.

Momblanch, A., L. Papadimitriou, S.K. Jain, A. Kulkarni, C.S.P. Ojha, A.J. Adeloye, and I.P. Holman. 2019. Untangling the water-food-energy-environment nexus for global change adaptation in a complex Himalayan water resource system. *Science Total Environment* 655: 35–47.

Mondal, S., and S. Mandal. 2019. Landslide susceptibility mapping of Darjeeling Himalaya, India using index of entropy (IOE) model. *Applied Geomatics* 11: 129–146.

Montemagni, C., C. Montomoli, S. Iaccarino, R. Carosi, A.K. Jain, H.-J. Massonne, and I.M. Villa. 2019. Dating protracted fault activities: Microstructures, microchemistry and geochronology of the Vaikrita Thrust, Main Central Thrust zone, Garhwal Himalaya, NW India. In *Crustal Architecture and Evolution of the Himalaya-Karakoram-Tibet Orogen*, eds. R. Sharma, I.M. Villa, and S. Kumar, 127–146. London: Geological Society, London, Special Publication 481.

Morrill, C., J.T. Overpeck, J. Cole, K. Liu, C. Shen, and L. Tang. 2006. Holocene variations in the Asian monsoon inferred from the geochemistry of lake sediments in central Tibet. *Quaternary Research* 65 (2): 232–243.

Mujtaba, S.A.I., R. Lal, H.S. Saini, P. Kumar, and N.C. Pant. 2018. Formation and breaching of two palaeolakes around Leh, Indus valley, during the late Quaternary. In *The Himalayan Cryosphere: Past and Present*, eds. N.C. Pant, R. Ravindra, D. Srivastava, and L.G. Thompson, 23–34. Geological Society, London, Special Publication 462.

Mukerjee, P., R. Sogani, N. Gurung, A. Rastogi, and K. Swiderska. 2018. *Smallholder Farming Systems in the Indian Himalayas: Key Trends and Innovations for Resilience.* London: International Institute for Environment and Development.

Murtaza, K.O. 2019. *Historical Changes in the Alpine Glacier System, A Case Study of Lidder valley.* Unpublished Doctoral dissertation. University of Kashmir, Kashmir (India).

Murtaza, K.O. and S.A. Romshoo. 2017. Recent glacier changes in the Kashmir Alpine Himalayas, India. *Geocarto International* 32(2): 188–205.

Murtaza, K.O., R.A. Dar, O.J. Paul, N.A. Bhat, and S.A. Romshoo. 2021. Glacial geomorphology and recent glacial recession of the Harmukh Range, NW Himalaya. *Quaternary International* 575–576: 236–248.

Mushtaq, F., A.G. Mantoo, A.S. Tirkey, and S.Z. Ahmad. 2022. Hazards in the perspective of Himalayan terrain: A review. In *Disaster Management in the Complex Himalayan Terrains Geography of the Physical Environment*, eds. S. Kanga, G. Meraj, M. Farooq, S.K. Singh, and M.S. Nathawat, 11–30. Springer.

Nag, D., B. Phartiyal, S. Agrawal, P. Kumar, R. Sharma, K. Kumar, A. Sharma, and M. Joshi. 2023. Westerly-monsoon variations since the last deglaciation from semi-arid Ladakh region, Trans Himalaya, India. *Paleogeography, Palaeoclimatology, Palaeoecology* 618: 111515.

Nagar, Y.C., A. Ganju, P.K. Satyawali, and N. Juyal. 2013. Preliminary optical chronology suggests significant advance in Nubra valley glaciers during the Last Glacial Maximum. *Current Science* 105 (1): 96–101.

Nakamura, A., Y. Yokoyama, H. Maemoku, H. Yagi, M. Okamura, H. Matsuoka, N. Miyake, T. Osada, D.P. Adhikari, V. Dangol, M. Ikehara, Y. Miyairi, and H. Matsuzaki. 2016. Weak monsoon event at 4.2 ka recorded in sediment from Lake Rara Himalayas. *Quaternary International* 397: 349–359.

Narsimha Rao, N., U. Devi, M.S. Shekhar, and G.P. Singh. 2021. Trends of winter precipitation extremes over Northwest Himalaya. *Hydrological Science Journal* 66 (13): 1882–1891.

Neckel, N., D. Loibl, and M. Rankl. 2017. Recent slowdown and thinning of debris-covered 502 glaciers in south-eastern Tibet. *Earth and Planetary Science Letters* 464: 95–102.

Negi, S.S. 2002. *Cold Deserts of India*, 2nd ed. New Delhi: Indus Publishing Company.

Negi, G.C.S. and P.P. Dhyani. 2012. *Glimpses of Forestry Research in the Indian Himalayan Region*. Almora, G.B. Pant Institute of Himalayan Environment and Development.

Negi, V.S., R.K. Maikhuri, D. Pharswan, S. Thakur, and P.P. Dyani. 2017. Climate change impacts in the western Himalaya: People's perception and adaptive strategies. *Journal of Mountain Science* 14 (2): 403–416.

Negi, V.S., R. Pathak, K.C. Sekar, R.S. Rawal, I.D. Bhatt, S.K. Nandi, and P.P. Dhyani. 2018. Traditional knowledge and biodiversity conservation: A case study from Byans valley in Kailash sacred landscape, India. *Journal of Environment Planning and Management* 61 (10): 1722–1743.

Nie, Y., Y. Sheng, L. Liu, S. Liu, Y. Zhang, and C. Song. 2017. A regional-scale assessment of Himalayan glacial lake changes using satellite observations from 1990 to 2015. *Remote Sensing Environment* 189: 1–13.

Nie, Y., Q. Liu, J. Wang, Y. Zhang, Y. Sheng, and S. Liu. 2018. An inventory of historical glacial lake outburst foods in the Himalayas based on remote sensing observations and geomorphological analysis. *Geomorphology* 308: 91–106.

Nirmal, B., K. Mohan, A. Tripati, B.A. Christensen, P.G. Mortyn, D. De Vleeschouwer, M. Prakasam, and K. Saravanan. 2023. Agulhas leakage extension and its influences South Atlantic surface water hydrography during the Pleistocene. *Palaeogeography, Palaeoclimatology, Palaeoecology* 615: 111447.

Nischal Attada, R. and K.M.R. Hunt. 2022. Evaluating winter precipitation over the Western Himalayas in a high-resolution Indian regional reanalysis using multisource climate datasets. *Journal of Applied Meteorology and Climatology* 61: 1613–1633.

NRSC. 2024. *GLOF Risk Assessment of Suru Glacial Lake in Indus River Basin*. National Remote Sensing Centre, Hyderabad.

Nuimura, T., A. Sakai, K. Taniguchi, H. Nagai, D. Lamsal, S. Tsutaki, A. Kozawa, Y. Hoshina, S. Takenaka, S. Omiya, K. Tsunematsu, P. Tshering, and K. Fujita. 2015. The GAMDAM glacier inventory: A quality-controlled inventory of Asian glaciers. *The Cryosphere* 9: 849–864.

Orr, E.N., L.A. Owen, S. Saha, M.W. Caffee, and M.K. Murari. 2018. Quaternary glaciation of the Lato Massif, Zanskar Range of the NW Himalaya. *Quaternary Science Reviews* 183: 140–156.

Osmaston, H. 1994. The geology, geomorphology and Quaternary history of Zangskar. In *Himalayan Buddhist Villages: Environment, Resources, Society and Religious Life of Zangskar, Ladakh,* eds. Crook, J. and H. Osmaston, 1–36. Delhi, Motilal Banarasidass Publishers.

Owen, L.A. 2009. Latest Pleistocene and Holocene glacier fluctuations in the Himalaya and Tibet. *Quaternary Science Reviews* 28: 2150–2164.

Owen, L.A., L.Y.N. Gualtieri, R.C. Finkel, M.W. Caffee, D.I. Benn, and M.C. Sharma. 2001. Cosmogenic radionuclide dating of glacial landforms in the Lahul Himalaya, northern India: Defining the timing of Late Quaternary glaciation. *Journal of Quaternary Science* 16: 555–563.

Owen, L.A., R.C. Finkel, and M.W. Caffee. 2002a. A note on the extent of glaciation throughout the Himalaya during the global Last Glacial Maximum. *Quaternary Science Reviews* 21 (1–3): 147–157.

Owen, L.A., U. Kamp, J.Q. Spencer, and K. Haserodt. 2002b. Timing and style of Late Quaternary glaciation in the eastern Hindu Kush, Chitral, northern Pakistan: A review and revision of the glacial chronology based on new optically stimulated luminescence dating. *Quaternary International* 97–98: 41–55.

Owen, L.A., M.W. Caffee, R.C. Finkel, and B.S. Seong. 2008. Quaternary glaciation of the Himalayan-Tibetan orogen. *Journal of Quaternary Science* 23: 513–532.

Owen, L.A., M.W. Caffee, K.R. Bovard, R.C. Finkel, and M.C. Sharma. 2006. Terrestrial cosmogenic nuclide surface exposure dating of the oldest glacial successions in the Himalaya orogen: Ladakh Range, northern India. *Bulletin Geological Society of America* 118: 383–392.

Palazzi, E., J. von Hardenberg, S. Terzago, and A. Provenzale. 2015. Precipitation in the Karakoram-Himalaya: A CMIP5 view. *Climate Dynamics* 45: 21–45.

Pandey, B.W., and A.S. Prasad. 2018. Slope vulnerability, mass wasting and hydrological hazards in Himalaya: A case study of Alaknanda Basin Uttarakhand. *Terræ Didatica* 14 (4): 395–404.

Pandey, A.C., S. Ghosh, and M.S. Nathawat. 2011. Evaluating patterns of temporal glacier changes in Greater Himalayan Range, Jammu & Kashmir India. *Geocarto International* 26 (4): 321–338.

Pandey, R., P. Kumar, K.M. Archie, A.K. Gupta, P.K. Joshi, D. Valente, and I. Petrosillo. 2018. Climate change adaptation in the western Himalayas: Household level perspectives on impacts and barriers. *Ecological Indicators* 84: 27–37.

Parajuli, A., A.P. Gautam, S.P. Sharma, K.B. Bhujel, G. Sharma, P.B. Thapa, B.S. Bist, and S. Poudel. 2020. Forest fire risk mapping using GIS and remote sensing in two major landscapes of Nepal. *Geomatics, Natural Hazards and Risk* 11 (1): 2569–2586.

Parshad, R., and P. Kumar, Snehmani P.K. Srivastava. 2019. Seismically induced snow avalanches at Nubra–Shyok region of Western Himalaya, India. *Natural Hazards* 99(5): 843–855.

Patel, L.K., P. Sharma, T.N. Fathima, and M. Thamban. 2018. Geospatial observations of topographical control over the glacier retreat, Miyar basin, Western Himalaya India. *Environmental Earth Sciences* 77 (5): 1–12.

Patel, L.K., P. Sharma, A. Singh, S. Oulkar, B. Pratap, and M. Thamban. 2021a. Influence of supraglacial debris thickness on thermal resistance of the glaciers of Chandra Basin, Western Himalaya. *Frontiers of Earth Sciences* 9: 706312.

Patel, A., A. Goswami, J.K. Dharpure, and M. Thamban. 2021b. Rainfall variability over the Indus, Ganga, and Brahmaputra river basins: A spatio- temporal characterisation. *Quaternary International* 575–576: 280–294.

Peng, M., X. Wang, G. Zhang, G. Veh, A. Sattar, W. Chen, and S. Allen. 2023. Cascading hazards from two recent glacial lake outburst floods in the Nyainqêntanglha range, Tibetan Plateau. *Journal of Hydrology* 626: 130155.

Perolo, P., M. Bakker, C. Gabbud, G. Moradi, C. Rennie, and S.N. Lane. 2019. Subglacial sediment production and snout marginal ice uplift during the late ablation season of a temperate valley glacier. *Earth Surface Processes and Landforms* 44 (5): 1117–1136.

Phadtare, N.R. 2000. Sharp decrease in summer monsoon strength 4000–3500cal yr B.P. in the central higher Himalaya of India based on pollen evidences from Alpine peat. *Quaternary Research* 53(1), 122–129.

Phartiyal, B., and A. Sharma. 2009. Soft-sediment deformation structures in the Late Quaternary sediments of Ladakh: Evidence for multiple phases of seismic tremors in the Northwestern Himalayan Region. *Journal of Asian Earth Sciences* 34 (6): 761–770.

Phartiyal, B., R. Singh, P. Joshi, and D. Nag. 2020. Late-Holocene climate record from a glacial lake in Ladakh range, trans-Himalaya India. *The Holocene* 30 (7): 1029–1042.

Phartiyal, B., R. Singh, D. Nag, A. Sharma, R. Agnihotri, V. Prasad, T. Yao, P. Yao, B. Karthick, P. Joshi, S.K.S. Gahlaud, and B. Thakur. 2021. Reconstructing climate variability during the last four millennia from trans-Himalaya (Ladakh-Karakoram, India) using multiple proxies. *Palaeogeography, Palaeoclimatology, Palaeoecology* 562: 110142.

Phillips, W.M., V.F. Sloan, J.F. Shroder Jr., P. Sharma, M.L. Clarke, and H.M. Rendell. 2000. Asynchronous glaciation at Nanga Parbat, northwestern Himalaya mountains, Pakistan. *Geology* 28: 431–434.

Phillips, E.R., D.J. Evans, J.J.M. van der Meer, and J.R. Lee. 2018. Microscale evidence of liquefaction and its potential triggers during soft-bed deformation within subglacial reaction tills. *Quaternary Science Reviews* 181: 123–143.

Pothula, S.K., and B.J. Adams. 2022. Community assembly in the wake of glacial retreat: A meta-analysis. *Global Change Biology* 28: 6973–6991.

Powell, R., T. Naish, R. Levy, and MIS Science Team. 2009. New records of the role of Antarctic ice sheets in late Cenozoic climate. *PAGES News* 17(1): 32–34.

Pradhan, S.P., S.D. Panda, A.R. Roul, and M. Thakur. 2019. Insights into the recent Kotropi landslide of August 2017, India: A geological investigation and slope stability analysis. *Landslides* 16: 1529–1537.

Pragya Kumar, M., A. Tiwari, S.I. Majid, S. Bhadwal, N. Sahu, N.K. Verma, D.K. Tripathi, and R. Avtar. 2023. Integrated spatial analysis of forest fire susceptibility in the Indian Western Himalayas (IWH) using Remote Sensing and GIS-based Fuzzy AHP approach. *Remote Sensing* 15: 4701

Prakash, S., R.K. Giri, and S. Chand. 2014. Environmental hazard-Landslides and avalanches (Kashmir region). *International Journal of Physics and Mathematical Sciences* 4: 87–99.

Prakash, S., M.C. Sharma, S. Sreekesh, P. Chand, V.J. Pandey, S.U. Latief, S. Deswal, I. Manna, S. Das, S.T. Mandal, and I.M. Bahuguna. 2022. Decadal terminus position changes and ice thickness measurement of Menthosa Glacier in Lahaul region of North-Western Himalaya. *Geocarto International* 37 (22): 6422–6441.

Prasad, S., A. Anoop, N. Riedel, S. Sarkar, P. Menzel, N. Basavaiah, R. Krishnan, D. Fuller, B. Plessen, B. Gaye, U. Röhl, H. Wilkes, D. Sachse, R. Sawant, M.G. Wiesner, and M. Stebich. 2014. Prolonged monsoon droughts and links to Indo-Pacific warm pool: A Holocene record from Lonar lake, central India. *Earth and Planetary Science Letters* 391: 171–182.

Prasath, R.A., A. Paul, and S. Singh. 2019. Earthquakes in the Garhwal Himalaya of the central seismic gap: A study of historical and present seismicity and their implications to the seismotectonics. *Pure Applied Geophysics* 176 (11): 4661–4685.

Pundir, S., V. Adlakha, S. Kumar, and S. Singhal. 2020. Closure of India-Asia collision margin along the Shyok Suture Zone in the eastern Karakoram: New geochemical and zircon U-Pb geochronological observations. *Geological Magazine* 157: 1451–1472.

Rafiq, M., and A.K. Mishra. 2018. A study of heavy snowfall in Kashmir, India in January 2017. *Weather* 74 (1): 15–17.

Rafiq, M., G. Meraj, A. Kesarkar, M. Farooq, S.K. Singh, S. Kanga, 2022. Hazard mitigation and climate change in the Himalayas-Policy and decision making. In *Disaster Management in the Complex Himalayan Terrains. Geography of the Physical Environment,* eds. S. Kanga, G. Meraj, M. Farooq, S.K. Singh, and M.S. Nathawat, 169–182. Springer.

Raina, V.K., and D. Srivastava. 2008. *Glacier Atlas of India.* Bangalore: Geological Society of India.

Ranhotra, P.S., J. Sharma, A. Bhattacharyya, N. Basavaiah, and K. Dutta. 2018. Late Pleistocene-Holocene vegetation and climate from the palaeolake sediments, Rukti valley, Kinnaur, Himachal Himalaya. *Quaternary International* 479: 79–89.

Rashid, I., and U. Majeed. 2018. Recent recession and potential future lake formation on Drang Drung glacier, Zanskar Himalaya, as assessed with earth observation data and glacier modelling. *Environmental Earth Sciences* 77: 429.

Rashid, I., S.A. Romshoo, and T. Abdullah. 2017. The recent deglaciation of Kolahoi valley in Kashmir Himalaya, India in response to the changing climate. *Journal of Asian Earth Sciences* 138: 38–50.

Rashid, I., U. Majeed, S. Aneaus, and M. Pelto. 2020. Linking the recent glacier retreat and depleting streamflow patterns with land system changes in Kashmir Himalaya, India. *Water* 12: 1168.

Rashid, I., K. Sain, J. Steiner, F. Ugalde, C.S. Watson, and M.J. Westoby. 2021. A massive rock and ice avalanche caused the 2021 disaster at Chamoli, Indian Himalaya. *Science* 373 (6552): 300–306.

Rasul, G., A. Saboor, P.C. Tiwari, A. Hussain, N. Ghosh, and G.B. Chettri. 2019. Food and nutrition security in the Hindu Kush Himalaya: Unique challenges and niche opportunities. In *The Hindu Kush Himalayan Assessment: Mountains, Climate change, Sustainability and People,* eds. P. Wester, A. Mishra, A. Mukherji, and A. Shrestha, 301–338. Springer.

Rather, M.I., I. Rashid, N. Shahi, K.O. Murtaza, K. Hassan, A.R. Yousuf, S.A. Romshoo, and I.Y. Shah. 2016. Massive land system changes impact water quality of the Jhelum River in Kashmir Himalaya. *Environmental Monitoring and Assessment* 188 (3): 185.

Rawat, M., S.K. Jain, R. Ahmed, and A.K. Lohani. 2023a. Glacial lake outburst food risk assessment using remote sensing and hydrodynamic modeling: A case study of Satluj basin, Western Himalayas, India. *Environmental Science and Pollution Research* 30: 41591–41608.

Rawat, M., R. Ahmed, S.K. Jain, A.K. Lohani, G. Rongali, and K.C. Tiwari. 2023b. Glacier-glacial lake changes and modeling glacial lake outburst flood in upper Ganga basin, India. *Modeling Earth Systems and Environment* 9: 507–526.

Raza, A., A. Razzaq, S.S. Mehmood, X. Zou, X. Zhang, Y. Lv, and J. Xu. 2019. Impact of climate change on crops adaptation and strategies to tackle its outcome: A review. *Plants* 8: 34.

Remya, S.N., T.H. Syed, A.V. Kulkarni, and R. Anand. 2022. Manifestation of topography and climate variations on long-term glacier changes in the Alaknanda Basin of Central Himalaya, India. *Geocarto International* 37 (26): 11010–11029.

Remya, S.N., T.H. Syed, A.V. Kulkarni, H.C. Nainwal, A. Mishra, and K.B.G. Raj. 2024. Observing changes in the present and paleo-glacial extents of major glaciers in the Alaknanda Basin of Central Himalaya. *Geomorphology* 461: 109287.

Richardson, S.D., and J.M. Reynolds. 2000. An overview of glacial hazards in the Himalayas. *Quaternary International* 65 (66): 31–47.

Rohling E.J., J. Yu, D. Heslop, G.L. Foster, B. Opdyke, and A.P. Roberts. 2021. Sea level and deep-sea temperature reconstructions suggest quasi-stable states and critical transitions over the past 40 million years. *Science Advances* 7: eabf5326.

Romshoo, S.A., S. Altaf, I. Rashid, and R.A. Dar. 2018. Climatic, geomorphic and anthropogenic drivers of the 2014 extreme flooding in the Jhelum basin of Kashmir, India. *Geomatics, Natural Hazards and Risk* 9 (1): 224–248.

Romshoo, S.A., B. Nabi, and R.A. Dar. 2024a. Influence of debris cover on the glacier melting in the Himalaya. *Cold Regions Science and Technology* 222 (10): 104204.

Romshoo, S.A., U. Ameen, M.H. Bhat, and T. Abdullah. 2024b. Comprehensive analysis of glacier recession (2000–2020) in the Nun-Kun Group of Glaciers, Northwestern Himalaya. *Journal of Mountain Science* 21 (3): 410–427.

Rounce, D.R., R. Hock, R.W. McNabb, R. Millan, C. Sommer, M.H. Braun, P. Malz, F. Maussion, J. Mouginot, T.C. Seehaus and D.E. Shean. 2021. Distributed global debris thickness estimates reveal debris significantly impacts glacier mass balance. *Geophysical Research Letters* 48: e2020GL091311.

Rounce, D.R., R. Hock, F. Maussion, R. Hugonnet, W. Kochtitzky, M. Huss, E. Berthier, D. Brinkerhoff, L. Compagno, L. Copland, D. Farinotti, B. Menounos, and R.W. McNabb. 2023. Global glacier change in the 21st century: Every increase in temperature matters. *Science* 379: 78–83.

Rowan, A.V., D.L. Egholm, D.J. Quincey, B. Hubbard, O. King, E.S. Miles, K.E. Miles, and J. Hornsey, 2021. The role of differential ablation and dynamic detachment in driving accelerating mass loss from a debris-covered Himalayan glacier. *Journal of Geophysical Research: Earth Surface* 126: e2020JF005761.

Rowold, D.J., D.P. Benedico, R. Gracia-Bertrand, S. Chennakrishnaiah, M. AlfonsoSanchez, F. Gaydeno, and R.J. Herrera. 2016. Ladakh, India: The land of high passes and genetic heterogeneity reveals a confluence of migrations. *European Journal of Human Genetics* 24: 442–449.

Roxy, M.K., C. Gananaseelan, A. Parekh, J.S. Chowdary, S. Singh, A. Modi, R.A. Kakatkar, S. Mohapatra, C. Dhara, S.C. Shenoi, and M. Rajeevan, 2020. Indian ocean warming. In *Assessment of Climate Change over the Indian Region,* eds. Krishnan, R., J. Sanjay, C. Gnanaseelan, M. Mujumdar, A. Kulkarni, and S. Chakraborty, 191–206. Springer.

Roy, P., T.R. Martha, N. Jain, and K.V. Kumar. 2018. Reactivation of minor scars to major landslides—A satellite-based analysis of Kotropi landslide (13 August 2017) in Himachal Pradesh, India. *Current Science* 115: 395.

Ruiz-Villanueva, V., S. Allen, M. Arora, N.K. Goel, and M. Stoffel. 2017. Recent catastrophic landslide lake outburst floods in the Himalayan mountain range. *Progress in Physical Geography* 41 (1): 3–28.

Sabeerali, C.T., and R.S. Ajayamohan. 2018. On the shortening of Indian summer monsoon season in a warming scenario. *Climate Dynamics* 50 (5–6): 1609–1624.

Sahu, R., and R.D. Gupta. 2020. Snow cover area analysis and its relation with climate variability in Chandra basin, Western Himalaya, during 2001–2017 using MODIS and ERA5 data. *Environmental Monitoring and Assessment* 192: 489.

Saif, B., M. Tahir, A. Sultan, M.T. Iqbal, T. Iqbal, M.A. Shah, and S. Gurmani. 2022. Triggering mechanisms of Gayari avalanche, Pakistan. *Natural Hazards* 112: 2361–2383.

Saktura, W.M., S. Buckman, A.P. Nutman, and V.C. Bennett. 2021. Late Jurassic Changmar Complex from the Shyok ophiolite, NW Himalaya: A prelude to the Ladakh Arc. *Geological Magazine* 158: 239–260.

Saktura, W.M., S. Buckman, A.P. Nutman, and R. Zhou. 2022. Jurassic-Cretaceous arc magmatism along the Shyok-Bangong Suture of NW Himalaya: Formation of the peri-Gondwana basement to the Ladakh Arc. *Journal of the Geological Society London* 179: 1–2.

Salerno, F., N. Guyennon, K. Yang, T.E. Shaw, C. Lin, N. Colombo, E. Romano, S. Gruber, T. Bolch, A. Alessandri, P. Cristofanelli, D. Putero, G. Diolaiuti, G. Tartari, G. Verza, S. Thakuri, G. Balsamo, E.S. Miles, and F. Pellicciotti. 2023. Local cooling and drying induced by Himalayan glaciers under global warming. *Nature Geoscience* 16: 1120–1127.

Sam, L., A. Bhardwaj, R. Kumar, M.F. Buchroithner, and F.J. Martín-Torres. 2018. Heterogeneity in topographic control on velocities of Western Himalayan glaciers. *Scientific Reports* 8: 12843.

Sangeeta and B.K. Maheshwari. 2019. Earthquake-induced landslide hazard assessment of Chamoli district, Uttarakhand using relative frequency ratio method. *Indian Geotechnical Journal* 49(1), 108–123.

Sangewar, C.V., and S.P. Shukla. 2001. Glacier inventory of Beas and Satluj basins in the districts of Kullu, Mandi, Rampur Bushar and Kinnaur, Himachal Pradesh. *Geological Survey of India Special Publication Series* 53: 257–264.

Sangewar, C.V., S.P. Shukla, and R.K. Singh. 2009. *Inventory of the Himalayan Glaciers.* Calcutta: Geological Survey of India.

Sardana, S., A. Verma, and A. Singh. 2019. Comparative analysis of rockmass characterization techniques for the stability prediction of road cut slopes along NH-44A, Mizoram, India. *Bulletin of Engineering Geology and the Environment* 78: 5977–5989.

Sati, V.P. 2024. Forest fires in the Indian Central Himalaya: Major drivers, implications, and mitigation measures. *Environmental Hazards* 23 (4): 390–404.

Sati, S.P., S.N. Ali, N. Rana, F. Bhattacharya, R. Bhushan, A.D. Shukla, Y. Sundriyal, and N. Juyal. 2014. Timing and extent of Holocene glaciations in the monsoon dominated Dunagiri valley (Bangni glacier), central Himalaya, India. *Journal of Asian Earth Sciences* 91: 125–136.

Sattar, A., A. Goswami, and A.V. Kulkarni. 2019. Hydrodynamic moraine-breach modeling and outburst flood routing-A hazard assessment of the South Lhonak lake, Sikkim. *Science of the Total Environment* 668: 362–378.

Saurabh, A., A. Thakur, and L. Kaur. 2020. A review on impact of climate change on apple production and scope of diversification in Himachal Pradesh. *International Journal of Current Microbiology and Applied Sciences* 9 (1): 1877–1884.

Savita, R.K. Luna, V.K. Dhawan, D. Kumar, A. Chaudhary, 2018. Impact of ban on green felling in Chirpine forests of Uttarakhand. *The Indian Forester* 144(8), 699–710.

Scherler, D., B. Bookhagen, M.R. Strecker, F. von Blanckenburg, and D. Rood. 2010. Timing and extend of late Quaternary glaciation in the western Himalaya constrained by 10Be moraine dating in Garhwal, India. *Quaternary Science Reviews* 29: 815–831.

Schildgen, T.F., P.A. van der Beek, M. D'Arcy, D. Roda-Boluda, E.N. Orr, and H. Wittmann. 2022. Quantifying drainage-divide migration from orographic rainfall over geologic timescales: Sierra de Aconquija, southern Central Andes. *Earth and Planetary Science Letters* 579: 117345.

Schmidt, S., and M. Nüsser. 2012. Changes of high altitude glaciers from 1969 to 2010 in the trans-Himalayan Kang Yatze Massif, Ladakh, northwest India. *Arctic, Antarctic and Alpine Research* 44 (1): 107–121.

Schmidt, S., M. Nüsser, R. Baghel, and J. Dame. 2020. Cryosphere hazards in Ladakh: The 2014 Gya glacial lake outburst flood and its implications for risk assessment. *Natural Hazards* 104: 2071–2095.

Schulz, H., U. von Rad, and H. Erlenkeuser. 1998. Correlation between Arabian Sea and Greenland climate oscillations of the past 110,000years. *Nature* 393 (6680): 54–57.

Schwanghart, W., A. Bernhardt, A. Stolle, P. Hoelzmann, B.R. Adhikari, C. Andermann, S. Tofelde, S. Merchel, G. Rugel, M. Fort, and O. Korup. 2016. Repeated catastrophic valley infill following medieval earthquakes in the Nepal Himalaya. *Science* 351: 147–150.

Shafiq, M.Ul, Z.Ul Islam, Abida, J.A. Tali, P. Ahmed, A.P. Dimri, and R. Mahmood. 2024. Assessment of climate variability and trends in different physiographic zones of North Western Himalaya. *Theoretical and Applied Climatology* 155(6): 5557–5576.

Shah, R.A., H. Achyuthan, A.M. Lone, S. Kumar, P. Kumar, R. Sharma, and C. Dash. 2020. Holocene palaeoenvironmental records from the high-altitude Wular lake, western Himalayas. *Holocene* 30 (5): 733–743.

Sharma, S., and A.D. Shukla. 2018. Factors governing the pattern of glacier advances since the Last glacial maxima in the transitional climate zone of the southern Zanskar ranges, NW Himalaya. *Quaternary Science Reviews* 201: 223–240.

Sharma, A., and B. Phartiyal. 2018. Late Quaternary palaeoclimate and contemporary moisture source to extreme NW India: A review on present understanding and future perspectives. *Frontiers in Earth Science* 6: 150.

Sharma, S., P. Chand, P. Bisht, A.D. Shukla, S.K. Bartarya, and Y.P. Sundriyal. 2016. Factors responsible for driving the glaciation in the Sarchu Plain, eastern Zanskar Himalaya, during the late Quaternary. *Journal of Quaternary Science* 31: 495–511.

Sharma, T., P. Semwal, S.S. Shah, H.C. Nainwal, and A. Mishra. 2018a. General geomorphological field observations around Satopanth glacier area, Garhwal Himalaya, Uttarakhand. *International Journal of Research and Analytical Reviews* 5 (4): 493–508.

Sharma, R.K., P. Pradhan, N.P. Sharma, and D.G. Shresta. 2018b. Remote sensing and in-situ based assessment of rapidly growing South Lhonak glacial lake in eastern Himalaya, India. *Natural Hazards* 93: 393–409.

Sharma, A., C. Wasko, and D.P. Lettenmaier. 2018c. If precipitation extremes are increasing, why aren't floods? *Water Resources Research* 54 (11): 8545–8551.

Sharma, M.C., I. Manna, and E. Chakraborty. 2022a. Contemporary dynamics and Holocene extent of glaciers in the Himalayas. In *Holocene Climate change and Environment,* eds. N. Kumaran and P. Damodara, 33–60. Cambridge, MA: Elsevier.

Sharma, S., S.P. Sati, N. Basavaiah, S. Pandey, Y.P. Sundriyal, N. Rana, P. Singh, S. Pradhan, A.D. Shukla, R. Bhushan, R. Bhatt, and N. Juyal. 2022b. Mid to late Holocene climate variability, forest fires and floods entwined with human occupation in the upper Ganga catchment India. *Quaternary Science Reviews* 293: 107725.

Shen, J., X. Liu, S. Wang, and R. Matsumotoet. 2005. Palaeoclimatic changes in the Qinghai lake area during the last 18,000 years. *Quaternary International* 136 (1): 131–140.

Shi, P., B. Han, K. Duan, L. Cao, A. Chen, and Y. Wu. 2023. Glaciers variation at 'shocking' pace in the northeastern margin of Tibetan Plateau from 1957 to 21st century: A case study of Qiyi glacier. *Atmosphere* 14 (4): 723.

Shrestha, B.B., and H. Nakagawa. 2016. Hazard assessment of the formation and failure of the Sunkoshi landslide dam in Nepal. *Natural Hazards* 82 (3): 2029–2049.

Shrestha, A.B., N.K. Agrawal, B. Alfthan, S.R. Bajracharya, J. Maréchal, and B. van Oort, (eds.). 2015. *The Himalayan Climate and Water Atlas: Impact of Climate on Water Resources in Five of Asia's Major River Basins*. ICIMOD: Kathmandu.

Shrestha, F., G. Xiao, N.R. Khanal, R.B. Shrestha, W. Li-Zong, P.K. Mool, and S.R. Bajracharya. 2017. Decadal glacial lake changes in the Koshi basin, central Himalaya, from 1977 to 2010, derived from Landsat satellite images. *Journal of Mountain Science* 14 (10): 1969–1984.

Shrestha, F., J.F. Steiner, R. Shrestha, Y. Dhungel, S.P. Joshi, S. Inglis, A. Ashraf, S. Wali, K.M. Walizada, and T. Zhang. 2023. A comprehensive and version-controlled database of glacial lake outburst floods in High Mountain Asia. *Earth System Science Data* 15: 3941–3961.

Shugar, D.H., M. Jacquemart, D. Shean, S. Bhushan, K. Upadhyay, A. Sattar, W. Schwanghart, S. McBride, M. Van Wyke Vries, M. Mergili, A. Emmer, C. Deschamps-Berger, M. McDonnell, R. Bhambri, S. Allen, E. Berthier, J.L. Carrivick, J.J. Clague, M. Dokukin, S.A. Dunning, H. Frey, S. Gascoin, U.K. Haritashya, C. Huggel, A. Kääb, J.S. Kargel, J.L. Kavanaugh, P. Lacroix, D. Petley, S. Rupper, M.F. Azam, S.J. Cook, A.P. Dimri, M. Eriksson, D. Farinotti, J. Fiddes, K.R. Gnyawali, S. Harrison, M. Jha, M. Koppes, A. Kumar, S. Leinss, U. Majeed, S. Mal, J. Muhuri, J. Noetzli, F. Paul, I. Rashid, K. Sain, J. Steiner, F. Ugalde, C.S. Watson, and M.J. Westoby. 2021. A massive rock and ice avalanche caused the 2021 disaster at Chamoli, Indian Himalaya. *Science* 373 (6552): 300–306.

Shukla, A., and P.K. Garg. 2019. Evolution of a debris-covered glacier in the western Himalaya during the last four decades (1971–2016): A multiparametric assessment using remote sensing and field observations. *Geomorphology* 341: 1–14.

Shukla, A., I. Ali, N. Hasan, and S.A. Romshoo. 2017. Dimensional changes in the Kolahai glacier from 1857 to 2014. *Environmental Monitoring and Assessment* 189: 5.

Shukla, T., V. Kumar, and M. Mehta. 2019. Response of the Himalayan glacial cycles to multiple equilibrium of climate system: A review. *Quaternary International* 507: 4–14.

Shukla, A.D., S. Sharma, N. Rana, P. Bisht, and N. Juyal. 2020. Optical chronology and climatic implication of glacial advances from the southern Ladakh Range, NW Himalaya, India. *Palaeogeography, Palaeoclimatology, Palaeoecology* 539: 109505.

Silva, P. Isabella, E. Garzanti, and M. Gaetani. 1992. Stratigraphy of the Chikkim and Fatu La formations in the Zangla and Zumlung units (Zanskar Range, India) with comparisons to the Thakkhola region (Central Nepal): Mid-Cretaceous evolution of the Indian passive margin. *Rivista Italiana Paleontologia E Stratigrafia* 97 (3–4): 511–564.

Siman-Tov, D., G.M. Stock, E.E. Brodsky, and J.C. White. 2017. The coating layer of glacial polish. *Geology* 45 (11): 987–990.

Singh, H., and S.K. Som. 2016. Earthquake triggered landslide- Indian scenario. *Journal of the Geological Society of India* 87 (1): 105–111.

Singh, K.K., and A. Singh. 2016. Detection of 2011 Sikkim earthquake-induced landslides using neuro-fuzzy classifier and digital elevation model. *Natural Hazards* 83 (2): 1027–1044.

Singh, G., and A. Pandey. 2024. Climate change induced disasters and highly vulnerable infrastructure in Uttarakhand, India: Current status and way forward towards resilience and long-term sustainability. *Sustainable and Resilient Infrastructure* 9 (2): 145–167.

Singh, A., A. Ganju, and S.K. Vaidya. 2004. Geohazard potential of snow avalanches in western Himalaya. *Bulletin of the Indian Geologists' Association* 37 (1&2): 15–23.

Singh, V., P.K. Thakur, V. Garg, and S.P. Aggarwal. 2018. Assessment of snow avalanche susceptibility of road network - A case study of Alaknanda basin. In *The International Archives of the Photogrammetry, Remote Sensing and Spatial Information Sciences, ISPRS TC V Mid-term Symposium Geospatial Technology - Pixel to People (Volume XLII-5)*, 461–468. Dehradun:India.

Singh, D., S. Ghosh, M.K. Roxy, and S. McDermid. 2019. Indian summer monsoon: Extreme events, historical changes, and role of anthropogenic forcings. *Wiley Interdisciplinary Reviews Climate Change* 10 (2): e571.

Singh, G., B.R. Nela, D. Bandyopadhyay, S. Mohanty, and A.V. Kulkarni. 2020. Discovering anomalous dynamics and disintegrating behaviour in glaciers of Chandra-Bhaga sub-basins, part of western Himalaya using DInSAR. *Remote Sensing of Environment* 246: 111885.

Singh, S., V. Sood, A.K. Taloor, S. Prashar, and R. Kaur. 2021a. Qualitative and quantitative analysis of topographically derived CVA algorithms using MODIS and 41 Landsat-8 data over Western Himalayas, India. *Quaternary International* 575–576: 85–95.

Singh, V., K.G. Misra, A.D. Singh, R.R. Yadav, and A.K. Yadava. 2021b. Little Ice Age revealed in tree-ring based precipitation record from the northwest Himalaya, India. *Geophysical Research Letters* 48: e2020GL091298.

Singh, L., M.J. Kaur, and D.K. Thakur. 2021c. Efficient fuelwood consumption with innovative solar water-heating system for forest conservation and mitigation of household carbon emission. *Current Science* 120 (5): 835–840.

Singh, S., A. Joshi, A, Sahu, R.A. Prasath, S. Sharma, and C.S. Dwivedi. 2022a. Himalayan landslides - Causes and evolution. In *Disaster Management in the Complex Himalayan Terrains - Natural Hazard Management, Methodologies and Policy Implications* Geography of the Physical Environment, eds. S. Kanga, G. Meraj, M., Farooq, S.K. Singh, and M.S.Nathawat, 33–42. Springer.

Singh, D., S. Kumar, and A. Chanjta. 2022b. Cultivation of fruit crops in Himachal Pradesh: Trend, concentration and diversification. *Journal of Asiatic Society of Mumbai* 96 (9)(I): 104–115.

Sivaprabha, S., I.M. Bhat, T. Ahmad, T. Tanaka, S. Balakrishnan, Y. Asahara, and D. Mukhopadhyay. 2022. Geochemical analysis of magmatic rocks from Shyok Suture Zone (SSZ) Trans-Himalaya, NW India: Insights for geodynamic evolution of the terrane. *Lithos* 410–411: 106594.

Soheb, M., A. Ramanathan, T. Angchuk, A. Mandal, N. Kumar, and S. Lotus. 2020. Mass-balance observation, reconstruction and sensitivity of Stok glacier, Ladakh region, India, between 1978 and 2019. *Journal of Glaciology* 66 (258): 627–642.

Sood, V., S. Singh, A.K. Taloor, S. Prashar, and R. Kaur. 2020. Monitoring and mapping of snow cover variability using topographically derived NDSI model over north Indian Himalayas during the period 2008–19. *Applied Computing and Geosciences* 8: 100040.

Souchez, R., G. VendenSchrick, R. Lorrain, and J.-L. Tison. 2000. Basal ice formation and deformation in central Greenland: A review of existing and new ice core data. In *Deformation of Glacial Materials,* eds. A.J. Maltman, B. Hubbard, and M.T. Hambrey, 13–22. Geological Society, London, Special Publication 176.

Space Applications Centre. 2016. *Monitoring Snow and Glaciers of Himalayan Region.* Ahmedabad, Space Applications Centre, Indian Space Research Organization.

Spicer, R.A. 2017. Tibet, the Himalaya, Asian monsoons and biodiversity—In what ways are they related? *Plant Diversity* 39: 233–244.

Sreejith, K.M., M.C.M. Jasir, P.S. Sunil, M.S. Rose, A.P. Saji, R. Agrawal, M.T. Bushair, K.V. Kumar, and N.M. Desai. 2024. Geodetic evidence for cascading landslide motion triggered by extreme rain events at Joshimath, NW Himalaya. *Geophysical Research Letters 51,* e2023GL106427.

Srivastava, D., C.V. Sangewar, M.K. Kaul, and K.S. Jamwal. 1999. Mass balance of Rulung glacier-A trans-Himalayan glacier, Indus basin, Ladakh. In *Proceedings of the Symposium on Snow, Ice and Glaciers—A Himalayan Perspective*, 41–46. Special Publication 53. Geological Survey of India.

Staubwasser, M., F. Sirocko, P.M. Grootes, and M. Segl, 2003. Climate change at the 4.2ka BP termination of the Indus valley civilization and Holocene south Asian monsoon variability. *Geophysical Research Letters* 30, 1425.

Storrar, R.D., M. Ewertowski, A. Tomczyk, I.D. Barr, S.J. Livingstone, A. Ruffell, B. Stoker, and D.J.A. Evans. 2019. Equifinality and preservation potential of complex eskers. *Boreas* 49: 211–231.

Su, B., D. Jiang, R. Zhang, P. Sepulchre, and G. Ramstein. 2018. Difference between the North Atlantic and Pacific meridional overturning circulation in response to the uplift of the Tibetan Plateau. *Climate of the Past* 14 (6): 751–762.

Sudan, F.K., R.K. Ganjoo, and S. Verma. 2022. Climate change and fuel wood energy use in cold desert of Drass in the Himalaya India. *Global Economics Science* 3 (2): 203–221.

Sultan, H., J. Zhan, W. Rashid, X. Chu, and E. Bohnett. 2022. Systematic review of multi-dimensional vulnerabilities in the Himalayas. *International Journal Environment Research Public Health* 19 (19): 12177.

Sun, Z., K. Yuan, X. Hou, K. Ji, C. Li, M. Wang, and J. Hou. 2020. Centennial-scale interplay between the Indian summer monsoon and the westerlies revealed from Ngamring Co, southern Tibetan plateau. *The Holocene* 30 (8): 1163–1173.

Svensson, A., K.K. Andersen, M. Bigler, H.B. Clausen, D. Dahl-Jensen, S.M. Davies, S.J. Johnsen, R. Muscheler, F. Parrenin, S.O. Rasmussen, R. Röthlisberger, I. Seierstad, J.P. Steffensen, and B.M. Vinther. 2008. A 60,000 year Greenland stratigraphic ice core chronology. *Climate of the Past* 4 (1): 47–57.

Szeliga, W., and R. Bilham. 2017. New constraints on the mechanism and rupture area for the M7.8 Kangra 1905 earthquake, NW Himalaya. *Bulletin of the Seismological Society of America* 107: 2467–2479.

Taloor, A.K., G.C. Kothyari, D.S. Manhas, H. Bisht, P. Mehta, M. Sharma, S. Mahajan, S. Roy, A.K. Singh, and S. Ali. 2021. Spatio-temporal changes in the Machoi glacier, Zanskar Himalaya, India using geospatial technology. *Quaternary Science Advances* 4: 100031.

Tamaddun, K.A., A. Kalra, M. Bernardez, and S. Ahmad. 2019. Effects of ENSO on temperature, precipitation, and potential evapotranspiration of North India's monsoon: An analysis of trend and entropy. *Water* 11: 189.

Tashi, T., and F.K. Sudan. 2022. A pragmatic approach to study vulnerability using livelihood vulnerability index: A case study from cold desert of Nubra valley, Ladakh (Himalaya). *Regional Economic Development Research* 3(2), 144–159.

Taylor, P.J., and W.A. Mitchell. 2000. The Quaternary glacial history of the Zanskar Range, north-west Indian Himalaya. *Quaternary International* 65 (66): 81–99.

Taylor, P.J., and W.A. Mitchell. 2002. Comment: Cosmogenic radionuclide dating of glacial land-forms in the Lahul Himalaya of northern India: Defining the timing of late Quaternary glaciation. *Journal of Quaternary Science* 17: 277–278.

Thakur, N., M. Rishi, T. Keesari, D.A. Sharma, and R. Singh. 2018. Assessment of changes in land use and land cover patterns using RS and GIS: A case study of upper Beas basin of Kullu valley in western Himalayas India. *Anusandhan* 14 (7): 1–10.

Thomson, S.I., and G.K. Vallis. 2018. Atmospheric response to SST anomalies. Part II: Background-state dependence, teleconnections, and local effects in summer. *Journal of Atmospheric Science* 75: 4125–4138.

Thompson, L.G., T. Yao, M.E. Davis, K.A. Henderson, E. Mosley-Thompson, P.N. Lin, J. Beer, H.A. Synal, J. Cole-Dai, and J.F. Bolzan. 1997. Tropical climate instability: The last glacial cycle from a Qinghai-Tibetan ice core. *Science* 276: 1821–1825.

Thompson, L.G., E. Mosley-Thompson, M.E. Davis, T.A. Mashiotta, K.A. Henderson, P.N. Lin, and Y. Tangdong. 2006. Ice core evidence for asynchronous glaciation on the Tibetan Plateau. *Quaternary International* 154–155: 3–10.

Thomson, J.R., P.B. Holden, P. Anand, N.R. Edwards, C.A. Porchier, and N.B.W. Harris. 2021. Tectonic and climatic drivers of Asian monsoon evolution. *Nature Communications* 12: 4022.

Tiwari, A., K. Sain, A. Kumar, J. Tiwari, A. Paul, N. Kumar, C. Haldar, S. Kumar, and C.P. Pandey. 2022. Potential seismic precursors and surficial dynamics of a deadly Himalayan disaster: An early warning approach. *Scientific Reports* 12: 3733.

Tripathy, A. 2011. *Exploring the History of India-Eurasia Collision and Subsequent Deformation in the Indus Basin, NW Indian Himalaya.* Doctoral degree. Arizona State University.

Tsering, K., M. Shrestha, K. Shakya, B. Bajracharya, M. Matin, J.L. Lozano, J. Nelson, T. Wangchuk, B. Parajuli, and M.A. Bhuyan. 2021. Verification of two hydrological models for real-time flood forecasting in the Hindu Kush Himalaya (HKH) region. *Natural Hazards* 110: 1821–1845.

Tu, X., G. Hu, X. Fu, Y. Zhang, J. Ma, Y. Wang, P.J.L. Gould, G. Du, H. Su, Z. Zhang, and J.W. Chapman. 2020. Mass windborne migrations extend the range of the migratory locust in East China. *Agricultural and Forest Entomology* 22: 41–49.

Turzewski, M.D., K.W. Huntington, and R.J. LeVeque. 2019. The geomorphic impact of outburst floods: Integrating observations and numerical simulations of the 2000 Yigong flood, Eastern Himalaya. *Journal of Geophysical Research - Earth Surface* 124: 1056–1079.

Vachula, R.S., J. Sae-Lim, and J.M. Russell. 2020. Sedimentary charcoal proxy records of fire in Alaskan tundra ecosystems. *Palaeogeography, Palaeoclimatology, Palaeoecology* 541: 109564.

Vannay, J.-C., B. Grasemann, M. Rahn, W. Frank, A. Carter, V. Baudraz, and M. Cosca. 2004. Miocene to Holocene exhumation of metamorphic crustal wedges in the NW Himalaya: Evidence for tectonic extrusion coupled to fluvial erosion. *Tectonics* 23(1): TC1014.

Verma, R.K., V. Pandro, S.N. Mishra, D. Raj, and A.J.K. Asaiya. 2019. Sal forest: A source of wild edible mushrooms for livelihood support to tribal people of Dindori district, Madhya Pradesh, India. *International Journal of Current Microbiology and Applied Sciences* 8 (1): 563–575.

Verma, S., M. Kumar, N. Ram, and P.S. Bisht. 2022. Changing patterns of agriculture in Uttarakhand state from 2010–2020. *IOSR Journal of Economics and Finance* 13(3): Ser. III, 1–10.

Verma, D., P. Govil, B. Kumar, and H. Khan. 2023. Variation in dynamics, controls and impacts of Agulhas leakage through Late Pleistocene: A Review. *Journal of Palaeosciences* 72: 43–54.

Viaggi, P. 2021. Quantitative impact of astronomical and sun-related cycles on the Pleistocene climate system from Antarctica records. *Quaternary Science Advances* 4: 10037.

Vijay, S., and M. Braun. 2018. Early 21st century spatially detailed elevation changes of Jammu and Kashmir glaciers (Karakoram–Himalaya). *Global Planetary Changes* 165: 137–146.

Walker, C.B., M.P. Searle, and D.J. Waters. 2001. An integrated tectono-thermal model for the evolution of the High Himalaya in western Zanskar with constraints from thermo-barometry and metamorphic modelling. *Tectonics* 20 (6): 810–813.

Wang, X., G. Pang, M. Yang, and G. Zhao. 2017. Evaluation of climate on the Tibetan plateau using era-interim reanalysis and gridded observations during the period 1979–2012. *Quaternary International* 444: 76–86.

Wang, Y., N. Wu, C. Kunze, R. Long, and M. Perlik. 2019. Drivers of change to mountain sustainability in the Hindu Kush Himalaya. In *The Hindu Kush Himalaya Assessment: Mountains, Climate change, Sustainability and People,* eds. P. Wester, A. Mishra, A. Mukherji, A.B. Shrestha, 17–55. Springer.

Wang, X., X. Guo, C. Yang, Q. Liu, J. Wei, Y. Zhang, S. Liu, Y. Zhang, Z. Jiang, and Z. Tang. 2020. Glacial lake inventory of high-mountain Asia in 1990 and 2018 derived from landsat images. *Earth System Science Data* 12: 2169–2182.

Wang, J., S. Yang, Z. Li, M. Lu, Z. Wang, and G. Wu. 2022. Optimal meridional positions of the Tibetan Plateau for intensifying the Asian Summer Monsoon. *Journal of Climate* 35: 3861–3875.

Wang, L., L. Wang, Y. Li, and J. Wang. 2023. A century-long analysis of global warming and earth temperature using a random walk with drift approach. *Decision Analytics Journal* 7: 100237.

Wang, X., G. Zhang, G. Veh, A. Sattar, W. Wang, S.K. Allen, T. Bolch, M. Peng, and F. Xu. 2024. Reconstructing glacial lake outburst floods in the Poiqu River basin, central Himalaya. *Geomorphology* 449: 109063.

Wani, S.A., F.A. Shaheen, and M.H. Wani. 2018. Cashmere producing smallholder nomads of Himalaya: Survival challenges of a system. *Small Ruminant Research* 163: 45–50.

Watanabe, Y., A. Abe-Ouchi, F. Saito, K. Kino, R. O'ishi, T. Ito, K. Kawamura, and C. Wing-Le. 2023. Astronomical forcing shaped the timing of early Pleistocene glacial cycles. *Nature* 4: 113.

Weiss, H., and R.S. Bradley. 2001. What drives societal collapse? *Science* 291: 609–610.

Weiskopf, S.R., M.A. Rubenstein, L.G. Crozier, S. Gaichas, R. Griffis, J.E. Halofsky, K.J.W. Hyde, T.L. Morelli, J.T. Morisette, R.C. Muñoz, A.J. Pershing, D.L. Peterson, R. Poudel, M.D. Staudinger, A.E. Sutton-Grier, L. Thompson, J. Vose, J.F. Weltzin, and K.P. Whyte. 2020. Climate change effects on biodiversity, ecosystems, ecosystem services, and natural resource management in the United States. *Science of the Total Environment* 733: 137782.

Wesnousky, S.G., Y. Kumahara, D. Chamlagain, I.K. Pierce, T. Reedy, S.J. Angster, and B. Giri. 2017. Large paleoearthquake timing and displacement near Damak in eastern Nepal on the Himalayan Frontal Thrust. *Geophysical Research Letters* 44 (16): 8219–8226.

Westoby, M.J., N.F. Glasser, J. Brasington, M.J. Hambrey, D.J. Quincey, and J.M. Reynolds. 2014. Modelling outburst floods from moraine-dammed glacial lakes. *Earth Science Reviews* 134: 137–159.

Wijngaard, R.R., A.F. Lutz, S. Nepal, S. Khanal, S. Pradhananga, A.B. Shrestha, and W.W. Immerzeel. 2017. Future changes in hydro-climatic extremes in the Upper Indus, Ganges, and Brahmaputra River basins. *PLoS One* 12 (12): e0190224.

Wirbel, A., A.H. Jarosch, and L. Nicholson. 2018. Modelling debris transport within glaciers by advection in a full-stokes ice flow model. *The Cryosphere* 12: 189–204.

Woodward, C.A., A. Slee, P. Gadd, A. Zawadzki, H. Hamze, A. Parmar, and D. Zahra. 2018. The role of earthquakes and climate in the formation of diamictic sediments in a New Zealand Mountain lake. *Quaternary International* 470(A): 130–147.

World Disasters Report. 2020. *Come heat or high water: Tackling the humanitarian impacts of the climate crisis together*. Geneva: International Federation of Red Cross and Red Crescent Societies.

World Disasters Report. 2022. *Trends in Disasters*. Geneva: International Federation of Red Cross and Red Crescent Societies.

Wu, N., S. Yi, S. Joshi, and N. Bisht, (eds.). 2016. *Yak on the Move: Transboundary Challenges and Opportunities for Yak Raising in a Changing Hindu Kush Himalayan Region*. ICIMOD: Kathmandu.

Wünnemann, B., D. Demske, P. Tarasov, B.S. Kotlia, C. Reinhardt, J. Bloemendal, B. Diekmann, K. Hartmann, J. Krois, F. Riedel, and N. Arya. 2010. Hydrological evolution during the last 15ky in the Tso Kar lake basin (Ladakh, India), derived from geomorphological, sedimentological and palynological records. *Quaternary Science Reviews* 29 (9–10): 1138–1155.

Xiao, J., S. Zhanga, J. Fana, R. Wena, Q. Xud, Y. Inouchie, and T. Nakamuraf. 2019. The 4.2ka event and its resulting cultural interruption in the Daihai lake basin at the East Asian summer monsoon margin. *Quaternary International* 527: 87–93.

Yadav, R.R., A.K. Gupta, B.S. Kotllia, V. Singh, K.G. Misra, and A.K. Yadav. 2017. Recent wetting and glacier expansion in the northwest Himalaya and Karakoram. *Scientific Reports* 7 (1): 1–8.

Yang, H., and Q. Wen. 2020. Investigating the role of the Tibetan Plateau in the formation of Atlantic Meridional Overturning Circulation. *Journal of Climate* 33: 3585–3601.

Yang, H., X. Shen, J. Yao, and Q. Weny. 2020. Portraying the impact of the Tibetan Plateau on global climate. *Journal of Climate* 33: 3565–3583.

Yang, W., J. Fang, and J. Liu-Zeng. 2021. Landslide-lake outburst floods accelerate downstream hillslope slippage. *Earth Surface Dynamics* 9: 1251–1262.

Yao, T., T. Bolch, D. Chen, J. Gao, W. Immerzeel, S. Piao, F. Su, L. Thompson, Y. Wada, L. Wang, T. Wang, G. Wu, B. Xu, W. Yang, G. Zhang, and P. Zhao. 2022. The imbalance of the Asian water tower. *Nature Earth and Environment* 3: 618–632.

Yao, T., Y. Xue, D. Chen, F. Chen, L. Thompson, P. Cui, T. Koike, W.K.-M. Lau, D. Lettenmaier, V. Mosbrugger, R. Zhang, B. Xu, J. Dozier, T. Gillespie, Y. Gu, S. Kang, S. Piao, S. Sugimoto, K. Ueno, L. Wang, W. Wang, F. Zhang, Y. Sheng, W. Guo, Ailikun, X. Yang, Y. Ma, S.S.P. Shen, Z. Su, F. Chen, S. Liang, Y. Liu, V.P. Singh, K. Yang, D. Yang, X. Zhao, Y. Qian, Y. Zhang, and Q. Li. 2019. Recent Third Pole's rapid warming accompanies cryospheric melt and water cycle intensification and interactions between monsoon and environment: Multidisciplinary approach with observations, modeling, and analysis. *Bulletin American Meteorological Society* 100: 423–449.

Yasuda, Y. and H. Tabata, 1988. Vegetation and climatic changes in Nepal Himalayas II. A preliminary study of the Holocene vegetational history in the Lake Rara national park area, west Nepal. *Proceedings of Indian National Science Academy* A 54(4), 538–548.

Yasunari, T. 2020. The uplift of the Himalaya-Tibetan plateau and human evolution: An overview on the connection among the tectonics, eco-climate system and human evolution during the Neogene through the Quaternary period. In *Himalayan Weather and Climate and Their Impact on the Environment,* eds. A.P. Dimri, B. Bookhage, M. Stoffel, and T. Yasunari, 281–305. Springer.

Yuan, M., T. Leirvik, and M. Wild. 2021. Global trends in downward surface solar radiation from spatial interpolated ground observations during 1961–2019. *Journal of Climate* 34 (23): 1–56.

Zaz, S.A., S.A. Romshoo, R.T. Krishnamoorthy, and Y. Viswanadhapalli. 2019. Analyses of temperature and precipitation in the Indian Jammu and Kashmir region for the 1980–2016 period: Implications for remote influence and extreme events. *Atmospheric Chemistry and Physics* 19: 15–37.

Zeitler, P.K., A.S. Meltzer, P.O. Koons, D. Craw, B. Hallet, C.P. Chamberlain, W.S.F. Kidd, S.K. Park, L. Seeber, M. Bishop, and J. Shroder. 2001. Erosion, Himalayan geodynamics and the geomorphology of metamorphism. *GSA Today* 11 (1): 4–9.

Zerboni, A., S. Biagetti, C. Lancelotti, and M. Madella. 2016. The end of the Holocene humid period in the central Sahara and Thar deserts: Societal collapses or new opportunities? *Past Global Changes* 24 (2): 60–61.

Zhang-Turpeinen, H., M. Kivim, H. Altonen, F. Berninger, E. Kister, K. Kister, O. Menyailo, A. Prokushkin, and J. Pumpanen. 2020. Wildfire effects on BVOC emissions from boreal forest floor on permafrost soil in Siberia. *Science of the Total Environment* 711: 134851.

Zharkova, V.V., S.J. Shepherd, S.I. Zharkov, and E. Popova. 2019. Oscillations of the baseline of solar magnetic field and solar irradiance on a millennial timescale. *Scientific. Reports* 9: 9197.

Zheng, P., J.B. Pedro, M. Jochum, S.O. Rasmussen, and Z. Lai. 2021a. Different trends in Antarctic temperature and atmospheric $CO_2$ during the last glacial. *Geophysical Research Letters* 48: e2021GL093868.

Zheng, G., S.K. Allen, A. Bao, J.A. Ballesteros-Cánovas, M. Huss, G. Zhang, J. Li, Y. Yuan, L. Jiang, T. Yu, W. Chen, and M. Stoffel. 2021b. Increasing risk of glacial lake outburst floods from future Third Pole deglaciation. *Nature Climate Change* 11: 411–417.

Zhou, Y., L.I. Zhiwei, and L.I. Jia. 2017. Slight glacier mass loss in the Karakoram region during the 1970s to 2000 revealed by KH-9 images and SRTM DEM. *Journal of Glaciology* 63: 331–342.

Zhou, R., J.C. Aitchison, K. Lokho, E.R. Sobel, Y. Feng, and J.-X. Zhao. 2020. Unroofing the Ladakh Batholith: Constraints from autochthonous molasse of the Indus Basin, NW Himalaya. *Journal of the Geological Society, London* 177 (4): 818–825.

Zhuang, M., A. Duan, R. Lu, P. Li, and J. Yao. 2022. Relative impacts of the orography and land-sea contrast over the IndoChina Peninsula on the Asian Summer Monsoon between early and late summer. *Journal of Climate* 35: 3037–3055.

# Index




R. K. Ganjoo and M. N. Koul, *The Himalayan Glaciers, Climate Change and Society*, https://doi.org/10.1007/978-3-031-78283-1

The manufacturer's authorised representative in the EU is Springer Nature Customer Service Centre GmbH, Europaplatz 3, 69115 Heidelberg, Germany. If you have any concerns regarding our products, please contact ProductSafety@springernature.com

Printed and bound by CPI Group (UK) Ltd, Croydon, CR0 4YY
30/09/2026
02225604-0018